SMITHSONIAN BOOK OF

GIANT PANDAS

SUSAN LUMPKIN
JOHN SEIDENSTICKER

SMITHSONIAN INSTITUTION PRESS • WASHINGTON AND LONDON

Project managers: Caroline Newman and
Vincent Burke
Designer: Linda McKnight
Production manager: Martha Sewall
Copy editor: Suzanne G. Fox
Production editor: Duke Johns

Library of Congress Cataloging-in-Publication
Data

Lumpkin, Susan.
Smithsonian book of giant pandas / Susan
Lumpkin and John Seidensticker.
p. cm.
Includes bibliographical references (p.)
ISBN 1-56098-038-4 (cloth : alk. paper)—
ISBN 1-58834-013-0 (pbk. : alk. paper)
1. Giant panda. 2. Wildlife conservation. I.
Seidensticker, John. II. Title.
QL737.C214 S45 2002
599.789—dc21 2001049282

British Library Cataloguing-in-Publication Data
is available

Manufactured in China, not at government
expense
09 08 07 06 05 04 03 02 5 4 3 2 1

♾ The paper used in this publication meets the
minimum requirements of the American
National Standard for Information Sciences—
Permanence of Paper for Printed Library
Materials ANSI Z39.48-1984.

The editors and designer would like to thank
Jessie Cohen, photographer, National
Zoological Park, for her contributions to this
work.

CONTENTS

Mei Xiang before leaving Wolong Reserve for the National Zoo.

FOREWORD

I will always remember my first meeting with Hsing-Hsing, the beloved male giant panda that lived at the Smithsonian National Zoological Park (SNZP). It was in 1995, and I had just started working at the zoo as a veterinarian. At the time, Hsing-Hsing was twenty-five years old and his arthritis was beginning to slow him down a bit. I stopped by the Panda House, found the panda keeper, and together we went into the off-exhibit holding area to see Hsing and feed him a piece of sweet potato. These visits, often early in the morning, became routine for me; I saw Hsing almost daily for the next four years, until he died of old age in November 1999. With the arrival of Mei Xiang and Tian Tian in December 2000, I have happily resumed this routine, although these youngsters get a few pieces of apple fiber biscuit instead of sweet potato.

Aside from fresh bamboo shoots and the occasional blueberry muffin, sweet potatoes were Hsing's favorite treat. The purpose of my visit was to check on the general health of my famous patient, but on the first visit I was mesmerized by his appearance and his behavior. He sat up on his haunches, paws on the wire mesh, and readily took a piece of cooked sweet potato into his mouth. The whole time he chewed gently, he looked at me quietly, his almond-shaped eyes surrounded by circles of black fur. His white face, round with huge chewing muscles, was perfectly framed by black ears that twitched back and forth as he ate. He was careful not to miss a morsel of the sweet potato, using his sixth digit (the panda's "thumb") to pick up a small piece that fell onto his other forearm. When he finished, he looked around to make sure there were no more treats, and then listened to his keeper, who directed him to go down into the yard and find his bamboo. Soon, but not before he was ready, he rambled down toward the willow tree out in front, where the public was waiting to see him. He sat up with his back against the tree, his white belly to the crowd and black legs splayed out in front of him. He picked up his bamboo and proceeded to eat his breakfast. The visitors were delighted.

I could watch Hsing eat bamboo for 30 minutes or more without even noticing the time. He would stretch, pick up a bamboo stalk, position it just so, select the leaves with his muzzle, and then delicately clip them with his tongue and teeth. At other times, he would take pawfuls and then chew the leaves in bunches, or break the stalk with his powerful jaws, strip off the outside layer, and eat the soft core. Tian Tian and Mei Xiang behave in the same ways. As I watch them, I wonder how an animal became so incredibly specialized to eat bamboo. How do they know which pieces of the plant to eat at any given time? What allows this large bear to live off plants instead of meat? How are we going to save this species, given how fast its habitat in China is vanishing? How can we secure the future of the wild giant panda in the face of natural bamboo die-offs and a shrinking habitat?

Despite the giant panda's popularity, these questions remain largely unanswered. We have much to learn about this incredible species, and we are racing against time. Over the past twenty-eight years, we learned a lot from Hsing-Hsing and his mate Ling-Ling. We initiated studies of behavior, biology, medicine, and reproduction; SNZP scientists exchanged information with colleagues in China and around the world. As Hsing and Ling entered their twilight years without successfully rearing their own offspring, breeding efforts were finally successful in China and Mexico, and, recently, in San Diego. Stabilizing the population of giant pandas in zoos and breeding centers is now an attainable goal. During the next ten years, we must learn how to save this species in the wild, and continue to work collaboratively to make this happen. Hsing-Hsing and Ling-Ling were known as ambassadors for peace. Tian Tian and Mei Xiang are known as ambassadors for conservation.

The SNZP is home to a wonderfully diverse collection of animals in addition to the giant pandas, ascending in size from ants to elephants. We attract three to four million visitors each year, welcome our guests free of charge, and transport them from the urban center of the nation's capital to the remote environments of elephants, bears, zebras, flamingos, Komodo dragons, and dozens of other species. Through our exhibits and our interpretive programs, we make every effort to convey to our visitors the very special bond between human beings and animals. Our goal is to strengthen and broaden that bond so we can work together to save the Earth's biodiversity and raise public awareness about endangered species. To do so, we offer genuinely innovative exhibits, which combine world-class facilities with science in action and educational outreach, and which promote conservation of the species on display. We are constantly working to ensure that our exhibits have the richness and depth of full-scale programs and that they encourage the animals to behave as they would in the wild.

The National Zoo's ten-year giant panda conservation, research, and training plan includes studies of nutrition, behavior, reproduction, habitat change, field ecology, and veterinary medicine. We are also offering to our Chinese colleagues professional training designed to put into practice what we learn. Our interpretive program and educational materials are meant to reach both our actual and our virtual Internet visitors. This book represents an important part of our effort to educate people about giant pandas and their world.

The giant panda exhibit itself is designed to wow our visitors through state-of-the art facilities, while also supporting on-site research and educational activities. Tian Tian and Mei Xiang live together during the day in spacious outdoor habitats outfitted so we can study their preferences for bamboo, rocks, water, trees, temperature, and humidity. We are also studying their behavior as they mature into adults. What we learn will be incorporated into plans for a new giant panda habitat.

And, of course, the animals themselves are the center of attention. Tian Tian and Mei Xiang are captivating, whether they are playing, eating, or sleeping. Hsing's engaging qualities are on display every day in these two pandas. In their first eight months at SNZP, Tian Tian and Mei Xiang attracted an amazing two million visitors.

As you read about giant pandas, please remember that they are ambassadors for many other endangered species whose long-term survival is equally uncertain. Perhaps the greatest legacy of the giant panda will be its ability to bring people and animals more closely together, so we can secure a future for all species.

Lucy Spelman, D.V.M.
Director
Smithsonian National
Zoological Park
Washington, D.C.

ACKNOWLEDGMENTS

The idea that there should be a new giant panda book emerged in April 2001. We submitted the final manuscript four months later, in August. In those months, we spent three weeks in May traveling in China, and nearly two weeks in June holed up on the southern coast of Spain writing the first drafts. At home, we abandoned kitchen and dining room to stacks and trails of books and papers, and pretty much ignored everything else to finish the text on deadline. But writing this book in the time allotted was possible only because we've been thinking about giant pandas and, more broadly, the conservation of large carnivores, for more than twenty years.

It is impossible to work at the Smithsonian National Zoological Park and not be involved at some level with giant pandas. John first explored giant panda reserves in China in early 1981, and published his first paper on them in 1983. Susan first saw giant pandas in Chinese zoos in late 1980; the following spring, she answered letters from people who had suggestions—white wine and candlelight was one—for encouraging the hapless Hsing-Hsing to mate with Ling-Ling. Much later, in late 1999, just as the National Zoo and Friends of the National Zoo (FONZ) were initiating efforts to bring another pair of giant pandas from China to the SNZP, we had a chance to visit China and giant panda reserves again, two of the same reserves John had seen before. This gave us a historical perspective on environmental change in China's panda landscapes. For the next year, we were immersed in the efforts to bring Mei Xiang and Tian Tian to the SNZP. Finally, in May 2001, we visited three new giant panda reserves in Sichuan. With SNZP Director Lucy Spelman and China Research Coordinator Mabel Lam, we were looking at how the funds the SNZP is contributing to giant panda conservation in China will be spent. These experiences shaped our thinking as we wrote this book.

We would like to thank Lucy Spelman and FONZ Executive Director Clinton A. Fields for supporting this project, and their deputies McKinley Hudson and James Schroeder for managing the business side of it.

We are grateful to World Wildlife Fund–U.S., which funded John's 1981 trip to China, as well as our 1999 visit. WWF's Ginette Hemley and Lu Zhi of the WWF–China Program arranged the 1999 trip; they, along with Peter Debrine, were also terrific traveling companions. Ginette, WWF giant panda specialist Karen Baragona, David Hellinga, John Mugaas, Blaire Van Valkenburgh, and several anonymous reviewers also provided many helpful comments.

Female Mei Xiang and male Tian Tian get to know one another.

The China Wildlife and Conservation Association, the American Zoo Association, the State Forestry Administration, and the Sichuan Forest Department hosted our 2001 trip "around the rim" of the Sichuan Basin. People from these organizations, Lu Xiaoping, Zhang Shanning, and Li Jianjou, accompanied us and looked after every detail superbly. They made the trip remarkably interesting and informative. The directors of the Qianfoshan, Tangjiahe, Meigu Da Feng Ding, and Yele reserves offered gracious hospitality. We would especially like to thank Zhang Liming of the Sichuan Forest Department, who answered hundreds of our questions about everything from the names of Chinese plants to the meaning of Chinese poems.

At the SNZP, Ben Beck, Lisa Stevens, Belinda Reser, Stuart Wells, Michael Davenport, and Alan Peters filled in for John while we were traveling and writing. At FONZ, Alex Hawes, Sue Zwicker, and Stephanie Elliott did the same for Susan. Alex also read and edited the entire manuscript in one marathon weekend. SNZP librarian Alvin Hutchinson tracked down obscure references and acquired dozens of papers and books in short order. Thanks to all of them.

Fujifilm USA generously donated slide and print film as well as a digital camera to our 2001 trip to China, just one part of this corporation's major commitment to giant panda conservation. Thanks to Judy Matson and Fran Bernstein for arranging the donation.

Caroline Newman and Vincent Burke of Smithsonian Institution Press did a great job coordinating this project. Linda McKnight created the book's beautiful design, and copy editor Suzanne Fox ensured that all of our mistakes were corrected. Thanks to all of them for their hard work.

We are especially grateful to Pam Leonard, Mark Elvin, and Ken Pomeranz for their help on our chapter on Chinese environmental history. Pam gave us a copy of her unpublished dissertation, dozens of other important papers, and an afternoon of her time talking to us about her experiences and sharing her ideas about people and the environment in China. Mark e-mailed page proofs of his forthcoming book, suggested further references, translated key sections of Wen Huanran's work on wildlife in Chinese history, and reviewed the chapter. Ken gave us permission to cite an unpublished paper, contributed new ideas and references, and also reviewed the chapter. We approached these individuals cold by e-mail and were astonished and gratified by their generous responses. Indeed, this was one of the pleasures of working on this book, which is much better for their contributions. Any errors, of course, are ours.

Finally, we thank Lesley Seidensticker, poet, dancer, artist, scholar, and our daughter. She puts up with us, and we love her for this and everything she is and does. This book is dedicated to her.

Susan Lumpkin
John Seidensticker

Following pages: Mei Xiang's and Tian Tian's first winter in Washington, D.C.

The giant panda under a pear tree seemed surprisingly small—smaller than Ling-Ling or Hsing-Hsing, the giant pandas that had so shaped our mental image of these animals. When we arrived at the farm, the young female panda was 30 feet away, sitting placidly under the tree, eating a pear. In just a moment, though, she turned, walked a few steps to a low stone wall, climbed effortlessly over it, and disappeared among the rumpled,

Previous pages and left: No sharp boundaries divide people and pandas in China. Cultivated slopes merge with higher-elevation bamboo forests. A panda leaving the forest inevitably meets people.

spent corn stalks, soggy with the early November morning dew. John followed, keeping his distance, as the panda ambled through the 7-foot-tall corn stalks on this steep slope deep in the Min Shan, in a place called Pingwu County, about 150 miles north of Chengdu in China's Sichuan Province.

The panda slipped in and out of view as John climbed up through the cornfield. Each corn stalk grew in its own soil-filled, rocked-up microterrace. This is very labor-intensive agriculture, but it has sustained mountain farmers for generations. The panda could maintain its effortless, robust, rolling gait for hours; and John was certain he couldn't match her pace for long on this slope. Then, after about 150 yards, the cornfield ended, and following her became impossible. Emerging from the field, the panda climbed the face of a sheer 15-foot slope and retreated into thick brush. There was no tracking the animal through that tangle.

The Min Shan, or Min Mountains, are the core of panda country. About half of the remaining wild giant pandas inhabit this range of skyscraping ridges and plunging valleys. But people live there, too—about 180 thousand in remote Pingwu County alone, mostly on small farms. The pear tree under which we first saw the panda grew amid a cornfield, a vegetable garden bright with ripe red chili peppers, and a small, two-story barn. The barn, a long, low-slung house, and a few other tidy outbuildings delicately placed among giant boulders were perched on a terrace overlooking a

narrow gorge. With its clear, rapid stream, the gorge would be spectacular in any setting, but here it was miniaturized by the soaring slopes, like a classic artistic rendering of a Chinese mountainscape.

END GAME, OR BEGINNING?

A giant panda sitting under a pear tree in a classically Chinese agrarian landscape goes to the heart of the challenges we face in securing the animal's future in the wild. Giant pandas are among the most endangered of animals; they are also among the best known. "The panda has been made the symbol of the fraternity of living things," George Schaller wrote in *The Last Panda,* his eloquent and disturbing story of the giant panda's plight. What a heavy burden that is, made heavier by diverse, sometime conflicting, visions of what this means.

Giant pandas are not deeply imbedded in the cultural history of China, unlike, for example, the tigers so richly represented in art, literature, and legend. China designated giant pandas a national treasure only in the 1960s. Killing a panda has provoked harsh penalties for decades, but through the 1980s and into the 1990s, poaching posed a major threat to wild pandas. Today, wild giant pandas live isolated in just a few mountain ranges in central China. From the mid-1970s to the mid-1980s, despite widespread concern for their conservation, giant pandas' total habitat was reduced by 50 percent to about 5,021 square miles. The Greater Yellowstone Ecosystem's area, by comparison, encompasses about 21,630 square miles. These mountains in China are a biodiversity hot spot, and thousands of species of plants and animals come under the giant panda's conservation umbrella.

Since the beginning of the twentieth century, the giant panda has been *the* trophy among animals, first for Western hunters, then for museums, and more recently for zoos. But the zoo population of giant pandas, inside and outside of China, is not yet self-sustaining, and there is unease among conservationists about giant pandas that are "rescued" and brought into breeding centers. We are all thrilled to see giant pandas romp and roll in zoos, and our desire to see more in zoos sometimes seems insatiable. Can this infatuation be translated into a future for wild giant pandas, or are we content just to have cuddly giant pandas where we can see them, divorced from their natural setting? Will we settle for giant pandas being the star attractions of theme parks? There is remarkable tension among people who want the giant panda to have a future, because different people value different outcomes.

Giant pandas in the wild appear self-effacing. Shielded by walls of bamboo, they hide while human activities erode their Chinese mountain homes. Human-caused environmental changes, if left unchecked, will lead to the panda's extinction. People have been altering these mountain habitats for millennia, perhaps for as long as there have been people in Asia, but the rate and extent of change now threatens to swamp the wildlife living there. The mountain habitats where giant pandas live are fragmented into disconnected patches of forest. Ecological and genetic processes have been disrupted. Something must be done. But in China, as in conservation activities everywhere, the people who are part of the problem must also be part of the solution.

The traditional approach to securing a future for an endangered species prescribed a simple formula: protect a few pieces of nature by keeping people out, and, left undisturbed, the animals and their habitat will survive there indefinitely. Conservation biologists now believe that this approach represents hospice ecology: watching carefully and compassionately as species inevitably slip into extinction.

Protected areas are important building blocks in wild giant panda conservation. Many of the remaining giant pandas live in the thirty-three or so protected areas scattered among the mountains. But these reserves are not environmental prisons. Many giant pandas live outside protected areas, move in and out of a protected area, or must travel between them. Recognizing this, conservationists are searching for ways to partner with people living among and near giant panda habitat, because they have the most to gain or lose.

Yale social scientist Stephen Kellert reminds us that ultimately, human values will determine whether we sustain giant panda landscapes. The giant panda's troubles stem from conflicts over values, over what matters to people. People will almost always put their own needs and the needs of

Left: Like these people harvesting cabbages in Wolong Reserve, most of China's people are farmers. Many of those who live in remaining panda habitat belong to ethnic minority groups.

This woman is a Chang.

their families and communities first, so somehow the presence of pandas must help, or at least not hinder, people's ability to meet their own needs. To secure a future for giant pandas, conservation actions must be adaptable, relevant, and made socially acceptable by linking their welfare to that of people who live near them. A better future for all of us lies in establishing sustainable relationships between people and resources.

Time and again, alarmists lament that our efforts on behalf of the giant panda are inadequate or inappropriate. Some believe that to acknowledge that some conservation progress is being made is seen as acquiescing, misguided, giving up the moral high ground. To be sure, if conditions do not change, we will soon lose wild giant pandas. But we like business leader Lou Gerstner's maxim: "No more prizes for predicting rain. Prizes only for building arks."

Conditions for wild giant pandas are difficult, but there are processes and programs in place to improve them. And many people are optimistic that these will eventually restore wild giant pandas and their habitats. Giant pandas can be stars in the ecological recovery that improves the lives of people who live near them. In many ways, China has made a commitment to the future of wild giant pandas, and so has much of the rest of the world. The giant pandas' fate is in our hands.

In their mountain habitat, pandas seek out sheltered, gentle terraces and avoid open, precipitous slopes.

2

ICONS AND AMBASSADORS

The Chengdu Research Base of Giant Panda Breeding covers about 90 acres on the outskirts of Sichuan's largest city. For crowded Chengdu, it is surprisingly spacious, but a planned expansion will increase its area more than five-fold, to about 575 acres. The base has well-equipped laboratories for sophisticated research on the reproduction, nutrition, endocrinology, and genetics of the world's best-loved bear. Among other projects,

scientists here, along with colleagues from the Smithsonian National Zoo and Zoo Atlanta, are perfecting techniques in artificial insemination and in-vitro fertilization, determined to find ways to boost the reproductive success of giant pandas in zoos.

As its name implies, the base is also a breeding center. When we visited with SNZP Director Lucy Spelman in May 2001, twenty-six giant pandas lived here in palatial indoor quarters with access to large outdoor habitats. Among them were four giant pandas born several months earlier, playful cubs that delighted us with their rough-and-tumble play. One mother is raising the four cubs, although only one is hers, so that the other mothers don't skip a year before producing a new baby—another method scientists are using to increase giant panda numbers.

Apart from its research function, the Research Base is lovely. Arching stands of bamboo shade the paths, and visitors stroll through its gardens of rhododendrons and peach trees to see giant as well as red pandas. Black-necked cranes stand around a large pond, and black swans glide through its still water.

Visitors also tour the Giant Panda Museum. Gracing this grand building is a monumental bronze statue of the animal the museum honors. It is striking how much the revered bear resembles a sitting Buddha, and the museum, a secular shrine. In hall after hall, graphics, images, skulls, skeletons, stuffed specimens, and pickled parts reveal all that is known about giant pandas, past and present.

Even the individual histories and genealogies of giant pandas in zoos throughout the world are portrayed, although the sign needs updating. The National Zoo's own Ling-Ling and Hsing-Hsing are represented here (as is former National Zoo scientist Devra Kleiman), but Mei Xiang and Tian Tian are not. Zhang Zhihe, our host at the Research Base, can, however, recite their histories.

Modern zoo biologists know the histories and genealogies of many of the other individual animals in their care, too, because this information is important for conservation breeding plans. But few think that the public would find such a chart—or dozens of pages of small print in dozens of books—very interesting, even for such popular species as lions and tigers and bears. Except, of course, for the black-and-white ones. People have an insatiable appetite for every morsel of the lives of giant pandas, and a profound desire to see the beasts for themselves.

Previous pages: In Chinese breeding centers, one super-mom might raise the babies of several other females along with her own.

Left: Combining conservations in the wild and in a zoo, the Chinese Research and Conservation Center for the Giant Panda is part of the Wolong Reserve.

The National Zoo's Ling-Ling achieved worldwide fame.

ANIMISM OR ANTHROPOMORPHISM?

One million people streamed through the National Zoo to see Mei Xiang and Tian Tian in the first four months after they were released from quarantine, including Bill Clinton on one of the last days of his presidency. The PandaCams on the Giant Panda Web site recorded about half a million visits in that same time, and thousands of words of welcome arrived by e-mail and letter. This was no surprise. Every zoo outside of China that has ever exhibited giant pandas has been similarly overwhelmed by people clamoring to see them.

And the animals themselves are treated like royalty—no other zoo animals have such lavish habitats or so much attention devoted to satisfying their every need, or what people believe they need. Airplanes materialize when a panda must be moved. The media cover their every action like paparazzi stalking Madonna or Brad Pitt.

People mourn when a giant panda they know dies, feeling the loss deeply. When Ling-Ling and, later, Hsing-Hsing died at the National Zoo, the media recounted their lives, the details of their deaths, and the effects of their absence on the human community. Ling-Ling died the same week as ballet legend Rudolf Nureyev. Both of these cultural icons were eulogized in print and on the air in roughly equal measure.

And not just zoo pandas. George Schaller described how his Chinese colleagues objected to the practice of simply giving numbers to the giant pandas that were radio-collared. He wrote, "Giving a name to a panda is a weighty matter, not something to be done casually or carelessly, for with a name the animal ceases to be just one among others in the forest and instead becomes a member of our community, one whose behavior we will observe, debate, and judge as much as our own."

Zoos strive to create great habitats to showcase their pandas.

It's as if giant pandas are people, too, or more so. Cynically, the four-star hotels and other deluxe enclaves frequented by Westerners in China are called "panda houses"; some Chinese resent the fact that zoo giant pandas may have it better than they do.

How did an animal with no particular historical significance in China, that no Westerner had ever heard of until 1869 nor seen alive until 1916, achieve the status of a demigod? What makes giant pandas the most charismatic of the megavertebrates?

Scholars have combed the arts and literature of China looking for evidence of an ancient affinity between people and pandas. This search has yielded just a few dozen written citations, some of them only dubious references to animals that might have been giant pandas. A common thread is that the animal is called some kind of black-and-white carnivore, such as a black-and-white fox or black-and-white leopard, that eats iron. Giant pandas occasionally go into villages to lick and chew on cooking vessels, accounting for this strange idea about the animal's diet. Sometimes, the text mentions that the animal eats bamboo, or lives in Sichuan, or that its fur is warm—panda pelts were widely used as beds and rugs.

The record in the arts is sparser still. While Chinese art is replete with bears, bamboo, and the misty mountain landscapes inhabited by giant pandas, no pandas are depicted until the twentieth century. One of China's earliest art critics (circa 847 C.E.) wrote, "Pictures contain the greatest treasures of the empire." This suggests that giant pandas were not considered "rare and precious animals" and "national treasures," as they are officially designated today.

To the extent that giant pandas had any significance to Chinese, it appears to have been confined to the early dynasties, the capital of which was Xi'an, in the northeast edge of the giant panda's range. An emperor of the Western Han dynasty (206 B.C.E.–24 C.E.) held a giant panda in his garden, where it was the most esteemed of the forty animals there. A Tang dynasty emperor (618–907 C.E.) reportedly sent two live giant pandas as a token of goodwill to Japan, perhaps the first instance of panda diplomacy.

Little examined is the role that giant pandas might play for the non-Han, mostly Tibetan peoples who have long lived in the giant panda's mountains and make up about 10 percent of Chinese citizens. In the late 1920s, hunters Kermit and Theodore Roosevelt (Junior) ventured into Yele, a land inhabited by a Tibetan people called the Yi. The Yi reportedly killed a giant panda only if it scavenged their beehives, and they could recall only once doing that, six years earlier. Otherwise, they thought of the bear as a minor god, and some local hunters were reluctant to help the Roosevelts in their hunt. Following the hunt, the local people refused to eat the meat. However, the Roosevelts also noted that people living about 25 miles from Yele, where there is now a panda reserve, had never heard of a giant panda. Today, ethnic Tibetan villagers are said to regard giant pandas with affection, and a panda that enters a village may be considered auspicious and thus tolerated.

In their 1966 book, *The Giant Panda,* Ramona and Desmond Morris cite a litany of reasons for the giant panda's cult status, based on a study of British children's reported attraction or aversion to various animals. First, some features of giant pandas make them resemble people: they have flat faces and no tails, they sit upright, their "thumbs" allow them to manipulate small objects, and their sex organs are hidden from view. Second, almost all of us gush and coo over babies, and many features of giant pandas appear infantile: the flat faces, disproportionately large eyes in disproportionately large heads, playfulness, apparent clumsiness, the appearance of softness, a round outline, short cute names, their need for protection. Biologist Stephen Jay Gould traced the evolution of Mickey Mouse from a slightly disreputable rodent to a lovable mouse as the character's appearance became progressively more infantile, more panda-like.

Giant pandas also win on the basis of being big but seemingly harmless to people and, with their bamboo diet, to other animals as well. The giant panda's striking black-and-white color appeals to people, too—think of other attractive black-and-white things: Dalmatians, piano keys, tuxedos, the bold black brush strokes of calligraphy on bright white paper, Oreo cookies. Then there is the panda mystique: it is rare, it lives like a hermit in fog-shrouded mountains, and its history of discovery is romantic,

The National Zoo's pandas attract thousands of fans on spring days in Washington, D.C.

full of drama and danger. And giant pandas are valuable, as the Morrises put it, "in terms of hard cash." People like that.

Finally, the Morrises suggest that the popularity of teddy bears prepared Westerners to love a "super teddy bear." The teddy bear first appeared in the United States, named after then-President Theodore Roosevelt. The story goes that in 1903 the president had been hunting all day without finding a bear to shoot. His hosts on the trip, not wanting him to go away disappointed, captured a black bear and tied it to a tree for him to kill. Upset by this unsportsmanlike conduct, the president refused to shoot it. The press reported the incident, and by 1904 the American people knew the story so well that Roosevelt changed his mascot from a moose to a bear for his re-election campaign. At the same time, the wife of a New York novelty store owner created toy bears and sold them under the name Teddy's Bear. People snapped them up, and soon retail giant Sears, Roebuck and Co. was selling the toys, which are still loved today.

This theory may be the weakest reason for the panda craze. Throughout history, bears have loomed large in the human psyche; giant pandas are particularly intriguing bears. This, as much as Teddy, may have helped

catapult giant pandas to stardom. Lots of people and things—from Martha Stewart to Beanie Babies—suddenly attract a large and devoted, sometimes fanatical, following for reasons that leave many of us scratching our heads. Since the Morrises wrote their book, social scientists have been studying this phenomenon, which has many of the characteristics of an epidemic. This research is eloquently summarized in Malcolm Gladwell's book, *The Tipping Point.* He asks, "Why is it that some ideas or behaviors or products start epidemics and others don't?" He then elucidates three agents of change—the Law of the Few, the Stickiness Factor, and the Power of Context. How do these apply to the giant panda craze?

Simply put, the Law of the Few holds that messengers matter; that a few people, by virtue of their social gifts and status, can help start an idea epidemic. Highly influential people, from princes to presidents, have participated in publicizing giant pandas. The Stickiness Factor says that the content of the message must be memorable, too, and giant pandas are certainly memorable. The Power of Context deals with the fact that epidemics are sensitive to the conditions and circumstances of time and place.

Pandas entered the consciousness of the West amid the turbulent times of the twentieth century. The baby panda Su Lin arrived in the United States in the dark days of the Depression, and received instant acclaim. A young giant panda named Ming arrived at the London Zoo at the end of 1938, just as war was breaking out in Europe, and died there at the end of 1944. During this terrible time, wrote the Morrises, "she became something of a symbol—a bit of fun in a funless, burning city. She was something exotically unreal in a epoch that was having its fill of reality." In 1972, Hsing-Hsing and Ling-Ling captured the rapt attention of a nation suffering from the disunity and trauma of the war in Vietnam. Most recently, the giant pandas that have come to the United States may be welcome, innocent distractions from our increasingly fractious politics and sometimes dangerous social milieu.

A PRIEST AND HIS BLACK-AND-WHITE BEAR

For both China and the West, giant pandas did not begin their meteoric rise to superstar status until about 130 years ago, when the French missionary Père Armand David sent the first specimen from Muping in western Sichuan, to the Paris Natural History Museum, and revealed this animal's existence to the Western world. In a description sent to the museum's director, he extolled the black-and-white bear, which he dubbed *Ursus melanoleuca,* as "easily the prettiest kind of animal I know," even though he never saw one alive. Père David's instant admiration presaged the widespread adulation to come.

Père David was among the first Westerners to penetrate the closed Middle Kingdom, which long had been hostile to foreigners. This was a time when the Western public closely

Left: This image of the National Zoo's Hsing-Hsing was once featured on a wine label.

Along with countless renditions of plush pandas (above), designers have fashioned panda likenesses into nearly every product imaginable, including backpacks (below) and toasters.

tracked the exploits of their envoys to the ends of the Earth. Any new find, let alone that of a beautiful, slightly unusual bear from a dangerous, distant land, attracted the attention of the public. But many years passed before the first spark truly caught fire. With only a handful of specimens in Western museums, mostly in Paris, and no new information coming from Westerners exploring China, the black-and-white bear, referred to in England as the parti-colored bear, remained virtually unknown.

The Morrises suggest that this all changed about 1900. A scientist at the British Museum examined a skull and assorted limb bones and concluded that they belonged not to a bear at all, but to a relative of the red panda. He renamed the parti-colored bear the great panda, and moved the museum's specimens out of the bear gallery. Other naturalists had earlier come to this erroneous conclusion, but for some reason, giving the rare and unusual animal a catchy new name created a buzz. Adventurers soon set their sights on securing more specimens.

Despite numerous Western scientific expeditions into giant panda country, some with broad botanical and zoological goals and others in single-minded pursuit of the giant panda, it wasn't until 1916 that the first Westerner saw one alive, and then another fourteen years before the next sighting. In the 1920s, when big-game hunters were celebrities, giant pandas became the Holy Grail—the ultimate trophy. Kermit and Theodore Roosevelt, the sons of the twenty-sixth president, took up the quest in 1929. Their China expedition was supported by the Field Museum in Chicago, which hoped to be the first in the United States to possess a giant panda specimen. After months of arduous travel through mountains rife with bandits, the brothers earned—with the invaluable help of the Yi people—the dubious distinction of being the first Westerners to shoot a giant panda.

The press glorified their exploit, and the American public was thrilled. Other museums sent expeditions, all wanting the prestige of displaying a stuffed giant panda. Still, by 1935, six Westerners had shot only four pandas between them (both Roosevelt brothers took credit for a single giant

Left: A bouquet of flowers and a child's touch create the illusion that pandas are like kindly uncles.

No matter what the setting, photographers clamor and jostle to capture pandas in pictures.

Artists put the finishing touches on a mural in the new home for Tian Tian and Mei Xiang at the National Zoo.

Right: People enjoy watching pandas in their zoo homes, with the mountains of Sichuan painted in the background, but pandas are more interested in each other.

panda, as did another pair of hunters). But the toll on giant pandas was far greater than four, because local hunters killed many more giant pandas to satisfy Western demand. The Smithsonian's Museum of Natural History, for instance, obtained fifteen skins and skeletons from an American missionary in Sichuan, who bought them from local hunters. Beginning with four animals killed for Père David, at least forty-two giant pandas left China as piles of skin and bones.

BRING 'EM BACK ALIVE

The next phase of panda mania began in the 1930s, when a couple of Americans vied to bring home a living giant panda. One was a famed hunter named Tangier Smith. His competitor was a New York socialite named Ruth Harkness, who picked up the torch after her giant-panda-seeking husband died in Shanghai. The tale of the conflict and competition between the hunter and the socialite is long and convoluted; fact and fiction are confused; both seemed to have some things to hide. But in the end, Ruth Harkness, by fair means or foul, captured the gold: she made it to San Francisco in 1936 with a baby giant panda named Su Lin. In San Francisco and later in Chicago and New York, huge crowds greeted them, clamoring to see the little cub that Harkness fed with a bottle and carried in her arms like a child. The press covered the story relentlessly, feeding the public frenzy.

Curiously, Harkness struggled to find a zoo willing to meet her price for the cub. The zoo in Chicago balked, as did New York's Bronx Zoo. Su Lin was offered to the National Zoo, too. Lucille Mann, wife of then-Director of the National Zoo William M. Mann, recalled a visit from Harkness and Su Lin: "We would have loved keeping it in Washington, but the astronomical price was far beyond the National Zoo's modest budget." Ultimately, Su Lin did find a home back at Chicago's Brookfield Zoo, which sponsored Harkness's further collecting trips to China.

Su Lin spawned an entire industry of giant panda memorabilia, from plush toys and jewelry to cartoons and books. Universal Studios, for instance, launched the animated cartoon character, Andy Panda, in 1939, and he went on to star in many more features until 1949; in comic books, Andy lived until 1962. In 1941, perhaps inspired by the popular appeal of Su Lin, the first *Curious George* story was published about a mischievous monkey found in Africa and taken to a New York zoo. Or perhaps people simply like the idea of rescuing wild animals. The immensely popular and long-running *Babar* series debuted in

1931, featuring a young African elephant brought to live in Paris, an experience that later qualifies him to become King of the Elephants.

Giant panda products continue to be a booming business. In the year of the arrival of Tian Tian and Mei Xiang, the National Zoo Stores carried more than 800 different panda items, from 30 cent postcards to $200 plush toys. In between were panda T-shirts, panda mousepads, and panda toasters that brown the bread in the shape of a panda.

Most important, Su Lin so enchanted Americans and Europeans that she is credited with ending the practice of killing giant pandas for science or for sport. The four Americans who had shot giant pandas set the example and vowed never to shoot one again. But ending the carnage had less to do with conservation than with an emotional response to a charming individual. Indeed, Su Lin's wild popularity spurred zoos' efforts to obtain more giant pandas for their adoring public, even though this was often as lethal to pandas as a shotgun. In 1939, for instance, Tangier Smith captured nine giant pandas, only five of which survived to reach London, where yet another soon died. There was concern in some quarters that giant pandas were disappearing in the wild because local hunters continued

to fill the markets of Chengdu with pandas both dead and alive to meet the perceived demand. One observer used this glutted market to claim that pandas were not rare at all. But zoo directors still wanted to hitch their star to this rising celebrity.

Even World War II only slowed the panda exodus from China. Between 1937 and 1946, a total of fourteen giant pandas arrived alive in Western zoos. No one knows how many died en route. And none lived very long by modern standards, or ever bred. Civil war, followed by the formation of the People's Republic of China in 1949, finally ended the drain. The last remaining giant panda in the West, and the last in a zoo anywhere, including China, died at the Brookfield Zoo in 1953. China allowed no exports again until a few animals went to Europe in 1957 and 1959, but quickly stopped them despite the continued pleas of European zoos. (With no diplomatic relations between the two nations, U.S. zoos could not even consider trading with China.) For the next fourteen years, China held fast to its giant pandas, releasing a few animals only to North Korea, a secretive nation from which no news emerged to the West.

GIANT PANDA DIPLOMACY

In a bold, unexpected move, U.S. President Richard M. Nixon met with People's Republic of China Chairman Mao Zedong in 1972 to mend the long rift between the two nations. This historic event was quickly overshadowed among the American public by some really big news: as a gesture of friendship, China promised to send a pair of giant pandas to the United States; to reciprocate, the United States would send a pair of Alaskan musk oxen named Milton and Matilda. Ling-Ling and Hsing-Hsing became the world's most famous diplomats. For months, American zoos lobbied for ownership of the exotic immigrants, their behavior reminding *Washington Post* writer Judith Martin of "a day care center with lots of aggressive children and only two teddy bears." The president finally decided to bestow them on the National Zoo, as befitting for a gift to the people of the United States. So Ling-Ling and Hsing-Hsing came to the National Zoo, and for the next twenty-six years reigned supreme as the only giant pandas in the United States. The giant panda cult thrived. Behind the scenes, the gift of giant pandas inaugurated a scientific collaboration between the National Zoo and Chinese biologists that remains strong.

China continued to send giant pandas as diplomats, to Tokyo, Paris, London, Mexico City, Madrid, and Berlin. With a growing awareness around the world of the wildlife conservation crisis, giant pandas also came to symbolize endangered species. It is telling that when Ling-Ling and Hsing-Hsing came to the National Zoo in 1972, no news account mentioned that they were endangered, even though the World Wildlife Fund (WWF) had adopted the giant panda as its logo in 1961. Today, no news account would omit this fact.

Left: Sitting upright like this, pandas resemble people.

Richard and Patricia Nixon brought the first giant pandas to the National Zoo in 1972.

Of course, it is always possible to love something too much. In the 1980s, as zoos everywhere sought to display giant pandas, China stopped giving them. Instead, short-term exhibition loans were made for which zoos paid $100,000 a month, money easily recovered with admission fees paid by throngs of people wanting to see the animals. Soon the situation was getting out of hand. Zoos as well as Disney World and the Michigan State Fair were clamoring to rent a panda. These traveling animals were precluded from breeding, so, apart from some questionable educational value, the loans did nothing to contribute to giant panda conservation, and may have even been detrimental to efforts to increase the zoo population. With giant pandas officially protected under the U.S. Endangered

Species Act in 1984, and listed on the Convention on International Trade in Endangered Species (CITES), the U.S. Fish and Wildlife Service (FWS) banned imports of giant pandas in 1988. Later, in 1990, the international conservation community also recommended a temporary ban on all giant panda loans.

Under intense political pressure, and amid lawsuits and the censure of the Columbus Zoo by the American Zoo Association (AZA), an import permit allowed that zoo to rent two male pandas in 1992, but only because the money derived from the loan was designated for the creation of a new giant panda reserve in China. The lawsuit was brought by the WWF and the AZA against the U.S. Department of the Interior, pressing it to review its giant panda loan policy and ensure that the loans complied with CITES and the Endangered Species Act. The WWF/AZA position was that all loans must not be detrimental to the pandas in the wild, must enhance the conservation of wild pandas, and must not be primarily for commercial purposes. A moratorium on panda loans went into effect in 1993 to allow the FWS to review the policy.

The moratorium was lifted only in 1998, when the San Diego Zoo was awarded a permit and welcomed Shi Shi and Bai Yun. This zoo had applied before the moratorium, so its permit was grandfathered, but it is complying with the terms of the new policy, at that time in draft form. A year later, in 1999, Zoo Atlanta received a permit to import Lun Lun and Yang Yang. And in 2000, the National Zoo submitted its permit application. Ling-Ling had died in 1992, Hsing-Hsing in 1999. For the first time in twenty-seven years, no giant panda lived in Washington, D.C. Despite the presence of giant pandas in California and Georgia, people across the country wanted giant pandas in the nation's capital. The FWS issued a permit in October 2000, and less than two months later, Mei Xiang and Tian Tian flew from Chengdu to Dulles International Airport outside of Washington, D.C.

The panda loans to Zoo Atlanta and the National Zoo, which were not without controversy, were approved only because each zoo had committed to long-term research and conservation programs designed to enhance the survival of giant pandas in the wild. Moreover, China had agreed to use the loan funds to support giant panda conservation.

The National Zoo's agreement includes a monetary contribution and research, conservation, and educational initiatives related to giant pandas in the wild and in zoos. The financial contribution is in two parts. First is the National Zoo's $1 million annual commitment to giant panda conservation in China. These funds are helping to support underdeveloped giant panda reserves. Second is an additional $300,000 or so a year for research at the National Zoo and in China, public education in China, and training of Chinese biologists, wildlife managers, veterinarians, and other specialists.

A major portion of the research program is devoted to increasing the number of giant pandas in zoos to create a self-sustaining population as insurance against the species' extinction in the wild. But the overarching goal is to ensure that the insurance never has to pay out.

Still other funds are supporting the public conservation education and exhibition programs based at the National Zoo. Tian Tian and Mei Xiang are ambassadors for giant pandas in the wild. These beautiful animals are helping to inspire people with a desire to save giant pandas—and wildlife and wildlands everywhere.

CONSERVATION IN CHINA

When the giant panda first emerged on the world scene, its range in China had been shrinking for centuries, apparently without comment. So while the Western obsession with giant pandas enhanced their symbolic and monetary value to the Chinese, it also created an incentive for their conservation and an awareness of their situation. The first evidence of this came in 1939, when the government of Sichuan briefly forbade the capture of giant pandas, perhaps the first wildlife conservation regulation in modern China. In 1946, a Chinese newspaper voiced fears that the market would drive giant pandas into extinction. The new People's Republic of China, established in 1949 when the Communist Party came to power, allowed no panda exports for nearly a decade, then explicitly addressed issues of giant panda conservation in 1962, when giant pandas were protected and hunting banned. In 1963, the first three giant panda reserves were established, growing to thirteen by 1989 and to thirty-three today, with many more planned in the near future. Setting aside land for these reserves came at great social and economic cost. For instance, in Tangjiahe Reserve, the operations of an entire logging commune were shut down and the farming commune that supported it resettled. A sign at the entrance to the reserve commemorates the 301 farmers who once lived on this site.

In 1941, two giant pandas en route to the United States, a gift from China in gratitude for war relief, stayed briefly in Chongqing in eastern Sichuan. They were visited by thousands of local residents, only a few of whom had ever seen a panda before. It wasn't until the 1950s that giant pandas first began appearing in Chinese zoos. In 1963, a pair at the Beijing Zoo produced the first zoo-born giant panda; soon after, in 1965, it was reported that a panda was born there following artificial insemination. Before American zoos were even thinking about conservation breeding programs, the Chinese were conducting one, and they continue to do so. The Chengdu Research Base, and the breeding facility at Wolong Giant Panda Reserve, testify to their commitment. Reproductive physiologists David Wildt, Jo Gayle Howard, and others, including many Chinese colleagues, have been working to improve breeding performance in China's captive pandas by, among other things, improving procedures for preserving semen and perfecting artificial insemination techniques.

John and two Smithsonian colleagues traveled to three giant panda reserves in 1981, becoming some of

Left: The chance to feed a panda a carrot is considered a special treat.

Zoo behaviorists provide pandas with interesting objects, such as this hay-filled burlap sack, to play with.

the first Westerners to visit these areas in nearly forty years. He wrote then that "From our long discussions a statement of the Chinese position might read: 'We have decided to increase the number of giant pandas and will do so.'" A Chinese proverb says, "Enough shovels of earth—a mountain. Enough pails of water—a river." Anyone who has seen the monumental ability of China to "just do it"—whether building a Great Wall or a Giant Buddha—will understand the confidence of the Chinese about their chances of saving giant pandas.

China succumbed to the lure of the giant panda cult far more readily than to other foreign influences. Père David went to China ready to convert its populace to Christianity; he left believing it would take forty or fifty *thousand* years to do so. Converting China and much of the rest of the world to the panda cult took just four or five decades. Traveling through Sichuan in 1999 and again in 2001, we saw giant pandas everywhere: on countless counters of plush pandas, on packages of cigarettes, on candy wrappers, on postage stamps, on tourist brochures, on billboards urging the residents of Chengdu to help improve the city's environment.

The giant panda cult has not been without critics. In 1939 in England, for instance, public frivolity and media frenzy surrounded a giant panda named Ming, which was much visited even by the royal family. But

Seemingly tender moments between pandas make it easy to succumb to the animals' charm.

Pandas are found in an increasing number of zoos around the world, but the goal must be increasing numbers of pandas in the wild in China.

with the outbreak of war with Germany looming ever larger, some people were appalled at the panda's dominance in the news. The Morrises cited this letter to the editor of London's *Daily Mail:* "[People] have rubber-necked this monstrosity until their eyes ached. The sickly sentimental panda plague has infected far more people than can ever hope to see it in the flesh. . . . And now the most nauseating of all symptoms of animal worship is making its appearance. The incorrigibly foolish are beginning to credit the panda with a soul."

Similar sentiments have been voiced by various commentators and letter writers ever since, but panda lovers dismiss them as curmudgeons. Other people object to the money spent to obtain giant pandas, citing the pressing need in Washington, D.C., for example, for better schools and safer streets that go underfunded, though it is unlikely that money spent on giant pandas would go to such priorities. Still others object to people fawning over giant pandas, and to scientists studying their every move, for the sake of the pandas' right to privacy.

After Hsing-Hsing died, *Washington Post* writer David Ignatius expressed his dismay at people's sentimental sense of loss at the death of an animal they really didn't know, compared to our concern for our fellow human beings. He wrote, "On that same front page [as the story reporting Hsing's death] was a grim story reporting that as many as 98 thousand Americans may die each year because of medical mistakes. But I suspect none of them will be memorialized with the same fervor as the Chinese bear."

In the summer of 2001, the first baby giant panda born at the San Diego Zoo was featured on the cover of *USA Weekend,* with the headline: "Beyond Cute: Why the San Diego Zoo's baby panda, Hua Mei, is the most important animal on the planet." She represents a successful zoo breeding program; a technological coup in that she was conceived via artificial insemination; and a conservation public relations triumph. As San Diego Zoo behaviorist Don Lindburg said, ". . . [seeing this infant] translates to a universal sense that pandas should always be a part of nature." Some skeptics question this conclusion. Do people who melt at panda babies want giant pandas to be part of wild nature in China, or simply part of their experience in local zoos?

There is also a deep human urge to want to protect appealing creatures like giant pandas. One youngster in Florida e-mailed Tian Tian and Mei Xiang, inviting them to find a home in her state, where there were no poachers and plenty of food. The idea that pandas are better off in good zoo homes than they are in the wild is prevalent, in the West as well as in China, where wild pandas are often "rescued" and sent to breeding centers. So, there is a danger that, with the giant panda population growing in zoos and breeding centers, less attention and money will be devoted to saving the giant panda and its habitat in the wild. Schaller described how many of his Chinese colleagues believed that giant pandas "would be happy living securely behind walls with a roof over their heads, good food, no enemies, no cares."

There is a Chinese fable about a fabulous sea bird that came to a town where it was feted by the local leader with wine and meats reserved for the most important events. But the bird couldn't eat these viands and died after three days. The leader treated the bird as he himself would have wanted to be treated, not as a bird should be treated. This fable may apply to saving giant pandas.

We have seen how the lust for giant pandas in the first half of the twentieth century and again during the rent-a-panda craze of the 1980s, left unabated, might have doomed the giant panda. Schaller, a vociferous critic of the recent competitive scramble for the scarce animals, wrote,

> There are not enough pandas, nor will there ever be enough, to provide animals for all those who clamor for them . . . . If we want to burden the panda with symbolism, reverence, and adulation, fine. However, we also have a moral obligation to maintain the species in the wild. With panda numbers dwindling year by year, not every zoo, not every country, can have them. The panda has not evolved to amuse humankind.

On the contrary, we hope that humankind has evolved to the point that we can save giant pandas, not just in panda houses, but in their forest homes.

3

IN DEEP TIME

One hundred degrees in the shade on that July day in 1981, on the western slope of Virginia's Blue Ridge. We pushed through the haze and humidity as we approached the radio-collared northern raccoon along an abandoned fencerow. From the signals emanating from her collar, we could pinpoint her, up in an old black locust tree. We moved cautiously until we could see her lying spread out, her legs dangling over both sides of a

Previous pages: Mei Xiang sprawls across a willow branch to cool off and sleep in the heat of the Washington, D.C., summer.

The versatile northern raccoon is equally good at keeping warm in the winter and cool in the summer.

Right: In contrast, it's much easier for a giant panda to stay warm in the winter than cool in the summer.

limb about 20 feet up. Hanging over the branch, she wasn't moving, but she was watching us as we approached, her sparkling eyes partially concealed by her black mask.

Scientists have linked raccoons and giant pandas in trying to understand the origins of these species deep in the past. Both members of the mammalian order Carnivora, the northern raccoon belongs to the family Procyonidae, the giant panda to the family Ursidae, the bears. These species have quite different morphological, physiological, and behavioral adaptations to their environments. Giant pandas now live in only a few mountain ranges in central China, where they feed almost entirely on bamboo. Northern raccoons, the quintessential American wildlife species, live nearly everywhere in the United States, where they feed omnivorously on everything from fruits and berries to crayfish and kitchen scraps.

Mammalogists call this raccoon the northern raccoon to distinguish it from its sister species, the tropics-living crab-eating raccoon. The two raccoon species look very similar but differ in their adaptations to temperate and tropical climates. We believe there were comparably similar giant panda forms: now-extinct tropical lowland giant pandas, and giant pandas that still live in temperate Chinese mountains. The northern raccoon's range has been steadily expanding, following the path of expanding human influence. The giant panda's mountain habitat has been steadily contracting, retreating in the face of intensifying human land use.

The northern raccoon's closest relatives—about twelve species that include crab-eating raccoons, coatis, kinkajous, olingos, and ringtails—live in the tropical and subtropical forests of South and Central America, but the fossil history indicates that this family had its origins in the tropical and subtropical forests of North America. These smallish carnivores, which range in weight from about 2 to 21 pounds, have a lower basal metabolic rate but average reproduc-

In addition to sun bears, North American black bears, and Asiatic black bears, the giant panda's relatives are (from left to right) the spectacled bear of South America; the brown bear of Eurasia and North America; the sloth bear of the Indian sub-continent; and the polar bear of the Arctic.

tive potential for the typical mammal of their size. Compared to other carnivores, their pelts provide modest to poor insulation and do not molt in a well-defined annual cycle. They have a modest to poor capacity for evaporative cooling, have no annual cycle of fattening and fasting, and only modestly diverse diets. Like their northern cousins, they are mostly nocturnal. In short, they adapted for living in the warmth of the tropics and subtropics.

Northern raccoons and, to a lesser extent, ringtails, are the only procyonid species to have escaped the tropics. The northern raccoon adapted to life in the temperate forests of North America, which are characterized by periods of feast and periods of famine. John Seidensticker and John Mugaas studied northern raccoons and found that the animals adapted to living in temperate forests by evolving several attributes. Northern raccoons are able to live in a variety of climates. They have good thermoregulatory abilities. They eat extraordinarily diverse diets. Their reproductive potential is high. The cornerstone of the northern raccoon's success was the evolution of a high basal metabolic rate, together with well-defined cyclic changes in body

fat and heat regulation, a high level of heat tolerance, and a high capacity for evaporative cooling.

Contrast this with the giant panda, with its uniform bamboo diet and low reproductive rate. Giant pandas do not have a high capacity for passive heat loss or evaporative cooling, nor a high level of heat tolerance; they probably have a low basal metabolic rate. As we watched the raccoon slumped over a branch on that hot summer day in 1981, we couldn't help but think about how the National Zoo's giant pandas Ling-Ling and Hsing-Hsing back in Washington, D.C., were faring. Their zookeepers had to bring them inside to the comfort of their air-conditioned dens when the temperatures rose enough to make them noticeably uncomfortable. So, while the northern raccoon can brave both the cold and the heat of North America, giant pandas are unable to beat the heat. These two species feel the world differently.

SEARCHING FOR ORIGINS OF DIVERSITY IN THE CARNIVORA

Although all the modern procyonids evolved in the New World, both the Procyonidae and the Ursidae had their origins in Eurasia. Scientists' efforts to understand the origins of the giant panda and its relationship to other Carnivora, especially to the bears and the procyonids, have become legendary. It is fascinating to trace how reasoning and evidence unfolded as more information and newer and more sophisticated techniques emerged to probe past evolutionary events. This is science at work: never static, always questioning, trying time and again for clarification.

The Harvard evolutionary biologist Ernst Mayr taught that the two great aspects of organic nature are life processes and diversity. This gives rise to such questions as, How do the morphological, physiological, and behavioral traits of organisms interact? What factors explain the diversity of life history adaptations, such as how species reproduce and disperse? Our understanding of the life processes and the nature and extent of the diversity in the bears is far from complete, but what we do know offers some insight into the conditions and processes that gave rise to unusual bamboo-eating bear species. Or maybe not so unusual. When we think "bear," the generalists such as brown bears and American black bears come immediately to mind. But other bear species, the ones less familiar to us, are far more specialized. The 220-pound sloth bear, native to the Indian subcontinent and Sri Lanka, is a specialist, feeding on mostly ants and termites. The 1,100-pound polar bear, native to the Arctic, hunts and kills mostly seals for a living.

The extinction of the dinosaurs about sixty-five million years ago was a major event in the history of the mammals, because it opened up space for other creatures to proliferate and inherit the Earth. In the ecospaces, or niches, that opened up, the order Carnivora arose from a member of a group of primitive carnivorans collectively called miacoids. (We use the term "carnivorans" to distinguish the Carnivora from the carnivorous mammals outside this order.) The miacoids first appeared in the fossil record about eighty to sixty million years ago. They were small, the size of a modern civet (about 2 to more than 6 pounds), with long bodies and tails, short flexible limbs, and small brains. These arboreal animals, with wide paws and spreading digits, lived in forests of the Northern Hemisphere. The miacoids differed from the creodonts, their sister taxa, in that the miacoids had evolved the key diagnostic feature of the Order Carnivora: a single pair of cutting teeth called the carnassials, which are the upper fourth premolar and the lower first molar. The carnassials gave the order its great evolutionary opportunity. By enhancing either the cutting or crushing aspects of these teeth, the carnivorans could become meat-eating specialists or plant-eating specialists. The creodonts, lacking this dental advantage, lost the competition and became extinct long ago.

Scientists divide the modern and extinct families of the Carnivora into two major groups. The cat-like group, called Feliformia, includes cats, hyenas, civets, and mongooses, and the extinct cat-like nimravids. The dog-like group, the Caniformia, includes bears, raccoons, weasels, skunks, badgers, dogs, and extinct bear-dogs. The aquatic carnivorans, including walruses, seals, and sea lions, are also linked to the Caniformia. This fundamental subdivision was complete by the late Eocene, about forty million years ago. In the last ten million years of the Eocene (forty-three to thirty-three million years ago), severe and rapid cooling shifted the Earth's climate from mostly warm and tropical toward cooler and seasonally arid. This is the era in the evolution of the carnivorans when, Blaire Van Valkenburgh says, they took on "a more modern aspect." New predators appeared in the form of the first canids, followed by the early weasels or mustelids and the cat-like nimravids.

"The key to understanding the history of carnivorous mammals over evolutionary time lies in the fundamental characteristic of the large predator adaptive zone that has changed little over the last 65 million years," wrote Van Valkenburgh, who, as a Johns Hopkins graduate student, joined us in radio-tracking raccoons from time to time and is now a distinguished professor at the University of California–Los Angeles. "There are a fairly limited number of ways to hunt, kill, and consume prey, and consequently, sympatric [living in the same area at the same time] predators have tended to diverge along the same lines, no matter where they lived. There are bone crackers, meat specialists, and omnivores." These ecospaces, or macroniches, have been and are occupied. But there is yet another ecospace, that of living as extreme plant-eating specialists. And a few carnivorans—the bamboo-specialist red pandas and giant pandas—evolved to occupy this ecospace. Their teeth reflect this specialization and differ from those of the omnivores or the pure meat-eaters.

George Schaller and his colleagues observed that "The animals [giant pandas] in our study areas had an ob-

Like most bears, and unlike giant pandas, brown bears have diverse diets that may include items from fruit and grubs to fish and moose calves.

vious predilection for the meat in our traps, consuming hair and skin as well as flesh. The panda's dentition, though apparently designed for crushing bamboo, is also well suited for crunching bones." Did the first pandas evolve as bone crackers and at some point, very early in their evolutionary history, get forced out of the bone-cracker niche by more efficient bone-cracking species? Was the bamboo-crunching niche open, and did the lineage survive as bamboo crunchers?

Similar species compete for food, and the more powerful species often kill their competitors, either directly or indirectly—tigers kill leopards and brown bears, pumas kill coyotes and bobcats, wolves kill coyotes and pumas. This key ecological process drives evolutionary trends among the carnivorans, as, over time, one of a pair of similar species replaces the other. Giant pandas are specialized for a diet of bamboo and thus can coexist with the omnivorous Asiatic black bear.

Through the interactions of its physiology, morphology, and behavior, the giant panda is locked into an extreme hypocarnivore, or bamboo-feeding, niche. They have evolved in ways that tie them to bamboo, a food resource that is relatively constant in its availability and nutrition level through the year. Asiatic black bears, on the other hand, occupy the omnivore niche: they exploit seasonally available plant and animal food resources. This is the very essence of the giant panda and its relationship to other bears.

A MOLECULAR PATH TO URSID ORIGINS

In the past, the road to understanding an animal's affinities traveled through the fossil record. A fossil trail to the origins of the giant panda exists and takes us back nearly to the place where they "budded" from other early bears, but this path is sometimes blocked by gaps. Another tack is to compare the anatomy of living animals, and infer relationships based on the similarity of anatomical features. There are pitfalls in this approach, too. It is not always clear whether structures are alike because the two animals are closely related, or because the structures evolved to serve similar purposes. Fortunately, there is another way of tracing evolutionary history. Recently, scientists have found that the echoes of a species' past, back to its origins, can be detected in its genes and chromosomes. Fossils remain essential clues that, coupled with powerful molecular tools, can solve many mysteries.

"The issue of the relationship of the giant panda to the other Carnivora, whether they should be considered in Ursidae and the Procyonidae or in their own family," John Eisenberg, the National Zoo's founding resident scientist, told us in 1981, "is a matter of how far back you go in defining where a family begins in the Carnivora radiation." Based on numerous lines of evidence, Eisenberg was a proponent of placing the giant panda in its own family, the Ailuropodidae, as first proposed by R. I. Pocock in 1921. Others draw the line elsewhere.

Left: A panda uses its thumb, actually an enlarged wrist bone, to maneuver bamboo stalks into position to be crushed between the broad molars in their powerful jaws.

A panda's enlarged wrist bone is an unclawed, thumblike digit on the forepaws.

In the National Cancer Institute laboratories of Stephen O'Brien, a longtime National Zoo research collaborator, Dianne Janczewski and Raoul Benveniste constructed a molecular tree, using a technique called hybridization of molecular DNA, that reveals that the Carnivora split into families tens of millions of years ago. Just when did the line that led to the giant panda diverge from the earliest bears? A long time ago, as Stephen O'Brien found. He and his colleagues applied four different techniques to tissue samples from modern members of the Carnivora to examine changes over time. Their results show that the progenitor of the subfamily Ailuropodinae, which gave rise to the giant

Young pandas often take refuge in trees to escape potential predators like dholes and leopards.

panda, diverged from the main bear line between twenty-five and eighteen million years ago. The progenitor of the spectacled bear, now confined to parts of South America, radiated between fifteen and twelve million years ago. In contrast, the radiation leading to the six remaining bear species occurred rather recently, between seven and five million years ago.

The six recent bear species share similar chromosome patterns. The giant panda's patterns, although divergent due to a chromosomal reorganization, is most like those of the other bears. The current convention is to include the giant panda in the subfamily Ailuropodinae in the family Ursidae. When these scientists examined the chromosome patterns of the Procyonidae, they found members of this group share many characteristics with the cats (Felidae) and other carnivore families, but not with bears.

DAWN BEARS

What do the fossils tell us? In the molecular approach to understanding the origins of a species, results are read on gels, sheets of film, and computer printouts. These techniques are accurate, and becoming more precise with each passing year. But we love the feel of a fossil tooth, or a leg bone, or, best of all, an entire skull. You can hold a fossil in your hands, feel the heft of it, look at it and know the kind of environment the animal it represents lived in, and wonder what changed, what event or megadisturbance led this species or group of species to diverge from others.

Paleontologists have used fossils to construct a phylogenetic framework for the Carnivora, as they have for many groups of vertebrates. Essentially a family tree for a group of species, a phylogeny shows the line or lines of direct descent in a group of organisms. It is also the study of the history of these relationships. The key bones for understanding relationships among carnivorans are the basicranium, the auditory bullae, teeth, and feet. Other mammals' bones, what is termed the postcranial skeleton, often reflect the animal's habitat, whether it was adapted for living in a forest, or a grassland, or in between. Teeth and jaws reflect mammals' diet, feet its gait and dexterity. All bears are plantigrade, meaning they walk on the soles of their hind feet, not on their toes, like cats do. Bears have five toes on each foot, with nonretractable claws. The giant panda has a modified wrist bone in its forepaw, usually called its thumb, that it uses to manipulate bamboo. The evolution of the carnassials, or meat-slicing teeth, separated the Carnivora from their ancestors. But in the bears, including giant pandas, these teeth have been secondarily flattened, indicating a shift from a cutting to a crushing function.

The auditory bulla is the part of the skull between the eardrum and the base of the outer ear. This bony chamber is filled with air and houses the ear ossicle of the middle ear. The auditory bulla is a resonating chamber, and the inflated auditory bulla that some species possess increases hearing sensitivity. Each is formed by

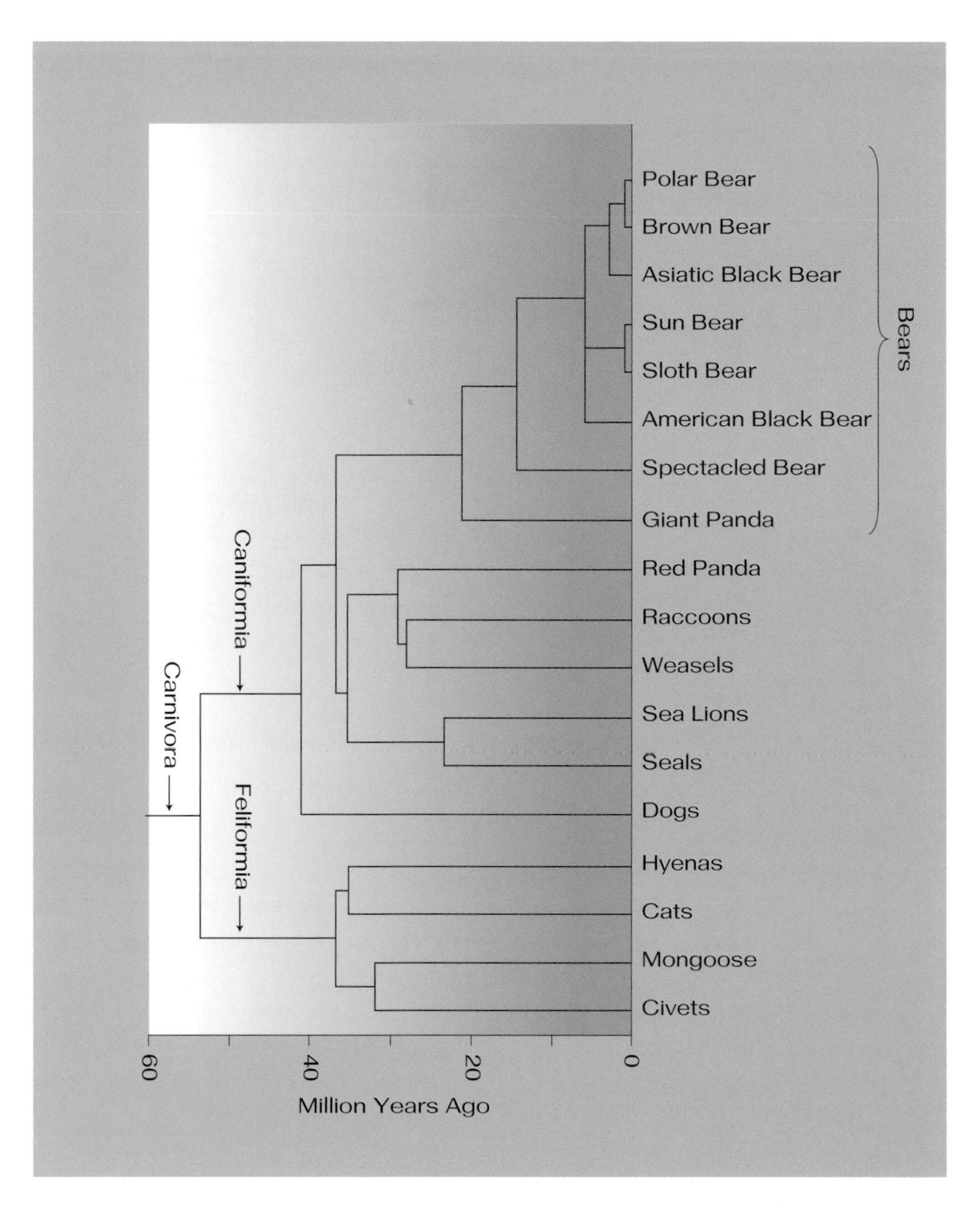

Combining morphological and genetic characteristics, scientists have created a tree that reflects relationships among the Carnivora.

the fusion of two bones: the entotympanic and the ectotympanic. The ear regions of the bears are unique among the carnivores. In most carnivoran families, the bulla is mostly formed from the entotympanic bone, with the ectotympanic forming just a small portion of the wall. In the bears, this is reversed, and you cannot even see the entotympanic bone. The cat-like carnivorans possess a unique arrangement in the auditory bulla. The bulla is divided by a septum that varies in depth in different species and enables them to hear with greater acuity the sound frequencies that match the calls of their primary prey. The bulla of the bear-like carnivorans are not divided by a septum. The giant panda's auditory bulla link them closely to the bears. "The tympanic cavity (in the giant panda) is remarkable for the small size of the bulla, but otherwise resembles that of *Ursus americana* [American black bear]," wrote Dwight Davis, the Field Museum scientist who undertook one of the most comprehensive comparative studies ever done on a mammal, using giant panda specimens in the Smithsonian Museum of Natural History and the Field Museum. Davis attributed the small size of the panda's bulla to adaptations in the skull necessary to accommodate the large powerful jaws needed for crushing bamboo.

Robert Hunt from the University of Nebraska State Museum reported that "The giant panda has had a long, separate history of evolution relative to the ursine bears. It descends as a distinct lineage (and is accorded its own subfamily Ailuropodinae) from

Present distribution
Historical records
Pleistocene fossils
Border of Yellow-Yangzi Lowlands
NEIMONGOL
LIAONING
HEBEI
Jinxi
Beijing
SHANXI
SHANDONG
NINGXIA
GANSU
Xi'an
Qinling Mountains
HENAN
JIANGSU
SHAANXI
Wanglang
Tangjiahe
Min Mountains
Pingwu
Qionglai Mountains
Qianfoshan
Wolong
Chengdu
Daxiongling Mountains
Meigudafengding
Yangzi River
Xiangling Mountains
Yele
Mianning
Xichang
Meigu
Liang Mountains
ANHUI
HUBEI
ZHEJIANG
JIANGXI
HUNAN
GUIZHOU
FUJIAN
Kunming
YUNNAN
GUANGDONG
GUANGXI
MYANMAR
VIETNAM
Lang Trang
LAOS
THAILAND
0
300 Mi.
0
500 Km.

the late Miocene protopanda, *Ailuarctos* [seven to eight million years ago]. The cheek teeth in *Ailuarctos* fossils, found in Lufeng, China, already began to resemble the cusps on the teeth that are so well developed in today's giant panda." The bears, excluding pandas, seem to be descended from *Cephalogale,* a fox-sized, running, meat-eater. From this ancestor came another raccoon-sized carnivore, *Ursavus,* named the dawn bear, thought to have given rise to non-panda living bears. This dawn bear gave rise to an early bear represented by the fossil *Agriotherium,* which were omnivores.

Agriotherium was a widespread genus of primitive bears that evolved in Europe during the Miocene and then spread to North America. *Agriotherium* gave rise to the *Plionarctos,* a much smaller primitive bear that first appeared in North America about seven million years ago. Probable descendants are the largest known land carnivores, the now-extinct long-footed *Arctodus simus,* the giant short-faced bear, and the short-footed *Tremarctos,* represented today by the South American spectacled bear. Both of these genera were widespread in North and South America. *Arctodus* have relatively long limbs, suggesting they could run faster and longer than the living bears. *Tremarctos* lived as far north as Florida seven to eight thousand years ago, but now is only found in northwest South America.

A second branch in the genus *Agriotherium* gave rise to the remaining bears we know today. The first fossil bear that is identifiable as the

While highly endangered, brown bears may still exist in Chinese mountain habitats like those in Wolong (previous pages).

genus *Ursus, U. minimus,* appeared in the Pliocene during a period of widespread, relatively rapid climatic change. There was a general increase in body size that would have improved the animals' temperature regulation in an increasingly seasonal climate. At the beginning of the Ice Ages about two and one-half million years ago, *Ursus minimus* gave rise to a much larger *Ursus etruscus,* which, in turn, gave rise to the present diversity of Northern Hemisphere bears—brown bear, polar bear, Asiatic black bear, American black bear, sun bear, and sloth bear, the subfamily Ursinae.

The major radiation in the Procyonidae occurred in the New World, with the stem procyonids coming in the early Miocene from Europe through Asia. Arboreal and omnivorous forms evolved within the tropical and subtropical climes of North America's central and middle latitudes. The first New World procyonid fossils have been dated at about eighteen million years old. Procyonids moved into South America about seven million years ago, and were the first placental carnivores to reach there. At this time, North and South America were separated by water. Paleontologists surmise that the animals rafted or island-hopped to reached South America, where the early procyonids faced a diverse assemblage of carnivorous marsupials, which the placental procyonids soon largely replaced. One procyonid form, *Chapalmalania,* emerged that resembled the giant panda in size and appearance. Was this a bamboo-eater, too? There is certainly plenty of bamboo in South America. The answer is lost in the swirling mists of time. This marvelous animal did not survive the turmoil of the Pliocene and Pleistocene.

RATCHETING INTO ECOSPACES

For three decades, Van Valkenburgh has been asking questions about why one group of carnivores displaced another in evolutionary time, or why one group suppressed the diversification of a replacement group, or clade, as scientists call such groups. Ecologists trying to understand the dynamics of an assemblage of living carnivores are constrained by the small scale and the short time frame in which they have to work. Van Valkenburgh has not been hampered by having to look at snapshots of the forces driving evolution. She has taken the long view: her time frame is about sixty-five million years. As a result, she has been able to study the response of one group or another over time spans long enough to encompass major environmental disturbances such as long droughts, sea-level changes, and mass extinctions.

Van Valkenburgh suggested that evolution in the Carnivora proceeds like a ratchet, ". . . by the loss of generalized features (small body mass, versatile dentition) as speciation tends to produce more specialized forms.

The very first panda-like carnivores, which lived seven to eight million years ago, have teeth suggesting that they were already specialized bamboo eaters.

Water is plentiful in panda habitat and pandas drink every day.

Given that evolutionary reversals are uncommon, clades that have become dominated by specialized forms (with larger body mass and dental specializations) are hypothesized to be at a disadvantage during environmental perturbations." The giant panda's subfamily Ailuropodinae designation represents an old and deep split within the family Ursidae, not a recent one. It appears that the giant panda ratcheted into an ecospace in which a steady supply of bamboo was the defining resource. And this ecospace has apparently been available for millions of years.

A driving force in evolution is intense competition and predation of one similar carnivore on another. In the giant panda's range, the ecospace defined by seasonally available plant and animal matter has been the domain of the large omnivores such as Asiatic black bears and brown bears. The specialized large carnivore meat-eating niche is the domain of dholes (Asiatic wild dogs), leopards, and, until recently, tigers. Giant pandas filled the only other ecospace available to a carnivoran. Biologist don't yet understand why the giant panda has not evolved a more efficient and specialized digestive system for processing bamboo. But there is no reason that the giant pandas will not continue to persist, unless they lose the bamboo forest in their mountain home through extensive modification by humans.

GIANT PANDAS IN A VIETNAM CAVE

Southwest of Hanoi, near the Laos border, lies the village of Lang Trang. Nearby, a limestone cave system protects a wealth of fossils dated to the middle Pleistocene, from about 150 thousand to about 500 thousand years ago. The caves and their fossils offer a window into climate and climate change and into the animals that lived through this period, animals such *Stegodon,* an extinct relative of the living elephants. And yes, the fossilized teeth of giant pandas are found here, along with the teeth of what were probably *Homo erectus,* a form very close to *Homo sapiens.*

Russell Ciochon of the University of Iowa and his colleagues explored these caves in cooperation with the Institute of Archaeology in Hanoi. Many species represented in the fossils occur in the region today: tigers, leopards, golden cats, sun bears, Asiatic black bears, palm civets, rhesus macaques, leaf-eating monkeys, gibbons, dholes, wild pigs, otters, elephants, sambar (a large deer), muntjac, gaur or wild cattle, water buffalo, tapirs, serow, Javan rhinos, and more. Species no longer in the area but represented in the cave fossils include giant pandas and orangutans.

Lang Trang is the southern edge of what paleontologists call the *Ailuropoda–Stegodon* fauna, which is also a typical assemblage of fossil mammals from this period in south China. This assemblage, including giant pandas, has also been unearthed at other sites in Vietnam, Thailand, and in Myanmar, but not farther south.

What were the climate and habitat conditions such that giant pandas' bones came to rest in a cave with those of an elephant relative, orangutans, and the rest? John Seidensticker and John Eisenberg studied the association between ungulates (hoofed mammals) and their habitat in southern Asia, and found they have strong habitat preferences. The species of ungulates in the caves thus suggests what the climate and habitat were like. Water buffalo and wild cattle prefer moist habitats but must have grasslands for grazing; these may be alluvial grassy plains, or grassy clearings maintained by human activities such as shifting cultivation or other disturbances. In southeast Asia today, there is an almost symbiotic relationship between wild cattle and wild water buffalo and people who clear land and then abandon it—the vegetation that subsequently grows in those clearings is what the cattle and water buffalo need.

Sambar and Asian elephants exist in a wide range of habitats, although elephants cannot live in arid environments. Greater one-horned rhinos are grazers confined to moist habitats with semi- to evergreen forest, usually in association with alluvial plains and tall grass; Javan and Sumatran rhinos are browsers and occur in rain forests. Wild pigs occupy diverse habitats but do not occur in dry, open areas. Serow (an Asian relative of the American mountain goat) live in steep, forested country with rock outcrops. Tapirs live in rain forest or in the gallery forest that occurs along streams in seasonal, monsoon forests.

Fossils of orangutans and pandas have been found in the same caves, suggesting there was once a tropical panda. Today's remaining pandas depend on bamboo, such as arrow bamboo (right), that grows in cool mountain habitats.

The purely meat-eating carnivorans, leopards, tigers, and dholes, specialize on larger ungulates and live where their prey do, focusing their activities in areas where prey are most abundant. And the highest numbers of ungulates occur where forest and meadow or alluvial plain converge to create a maximum interdigitation of cover types.

What of the other species in the cave? Gibbons live in rain forests and orangutans reach their highest numbers along rivers that wind through rain forests. Macaques use a variety of forest types but almost always live near water. Fossil otter teeth were present. Because Lang Trang is along a stream, the presence of species associated with rivers and streams and gallery forests is expected. Sun bears are omnivores that live in rain forests. Asiatic black bears are omnivores that live in a variety of forest types, especially hilly seasonal forests, and not usually rain forests. And giant pandas persist where there are extensive stands of bamboo.

There are no species in the Long Trang cave that use semi-arid habitat with low shrubs or moderate grass cover, although fossils from horses and from bovids, such as gazelles, indicate these open habitats occurred across northern China. The animals that were present at Lang Trang tell us that the habitat comprised mixed forest types with what were probably disturbed areas that supported grassy meadows and extensive stands of bamboo. It was also warm, indicating that the resident panda was the tropical lowland form.

A GIANT APE IN THE MIX

In thinking about how this Pleistocene faunal assemblage at Lang Trang might help us to understand the living giant panda, it is useful to consider how climatic changes molded this landscape. Here and elsewhere in southeast Asia, the vegetation alternated between rain forest, seasonal forest, and drier savannas and grasslands. Also, many large mammals became extinct in the Pleistocene, probably due to abrupt climate change and human hunting and clearing. The earliest hominid occupation of southeastern China and northern mainland southeast Asia probably occurred between one million and one-and-one-half million years ago. In Vietnam, the latest deposits containing orangutans are from twenty-three thousand years ago, and those containing giant pandas are from eighteen thousand years ago, a cool period with greatly reduced sea levels.

It is fascinating that the teeth of the largest-ever primate, the giant ape *Gigantopithecus,* have been found at sites containing fossil teeth of both giant pandas and *Homo* species. Comparing the jaw of a *Gigantopithecus* and a gorilla's jaw is like comparing the jaw of a giant panda and a grizzly's jaw. The big, extinct ape and the giant panda jaws evolved for the heavy grinding of very tough plant material. Both jaws are thicker, deeper, and more massive than those of their counterparts. Smithsonian paleobotanist Dolores Piperno examined some *Gigantopithecus* teeth and detected phytoliths, macroscopic pieces of silica found in many plants. Some of these phytoliths were from bamboo, some possibly from the fruits of durian or jackfruit. Russell Ciochon, the scientist-sleuth who unearthed the Lang Trang cave fossils, tentatively concluded that *Gigantopothicus* was primarily a bamboo eater. There were no *Gigantopithecus* fossils at Lang Trang, but they have been found in other Vietnam cave sites to the north and in China.

Adult male *Gigantopithecus* were estimated to have weighed as much as 1,200 pounds and stood 10 feet tall. Females were much smaller. For comparison, a large adult male gorilla stands just over 5½ feet and weighs 400 pounds. *Giganotopithecus* became extinct near the end of the middle Pleistocene. Were they victims of hunting by early humans as well as of the shift toward drier conditions that re-

duced their moist forest strongholds? Does this also account for the demise of the *Stegodon,* and the disappearance of orangutans from mainland southeast Asia? And is this why there are no tropical giant pandas? It seems a reasonable working hypothesis.

BAMBOO AND PLEISTOCENE LANDSCAPE CHANGES

It boils down to a simple axiom: to understand the giant panda, follow the bamboo. The bamboo that pandas eat thrives under humid conditions where there is annual precipitation of at least 60 inches. It doesn't grow in arid habitats or environments with seasonally low humidity. It doesn't grow in dry valley bottoms or on slopes desiccated by cold winds. Where climatic conditions are favorable, however, bamboo responds favorably and abundantly to disturbance. Indeed, there are extensive stands of bamboo in disturbed southeast Asian rain forests and monsoon forests, but bamboo is relatively scarce in undisturbed tropical and subtropical forests.

Pleistocene environments in Asia were in great flux, with fluctuating temperatures, levels of precipitation, and sea levels. Tree lines and vegetation zones shifted as much as 4,900 feet. Humid warm periods, supporting vegetation similar to modern subtropical rain forest, prevailed in south China from 140 to 240 thousand years ago and again from 100 to 130 thousand years ago. During these periods, warm-adapted primates, for example, moved north.

The tropical giant panda is extinct. It is up to us to save the cold-adapted mountain form.

During colder, drier periods, they moved south. However, during the dry periods, there were probably tropical species trapped in spots, called refugia, where tropical conditions prevailed. Lang Trang may be such a refugia.

Fossil giant pandas as part of the *Stegodon–Ailuropoda* assemblage have been discovered in a wide range of sites in south China, as well as north as far as Beijing and as far south as Lang Trang in Vietnam. Clearly, when this fossil association included orangutans and gibbons, the forest type was tropical. But these species were not always part of the mix, and at those times the association could indicate cooler environments. Giant panda fossils from about the middle Pleistocene onward are placed with today's giant pandas, *Ailuropoda melanoleuca,* while earlier ones are considered a different species, *Ailuropoda microta,* an animal about half the size of today's giant panda with a slightly different dentition. It is interesting that small *Ailuropoda microta* is found in association with the giant ape *Gigantopithecus* at one of the southern *Stegodon–Ailuropoda* sites. It is possible that their difference in size let the two bamboo-eaters occupy the same ecospace.

There are many reasons why a species has different size morphs, or types, such a big and a small panda. Bergmann's rule states that ". . . races from cooler climates, in species of warm-blooded vertebrates, tend to be larger than races of the same species living in warmer climates." Large animals better tolerate cold temperatures than small ones; they can also go without food longer. It's no coincidence that the smallest northern raccoons are from the subtropical Florida Keys and are half the size of the biggest northern raccoons living in Minnesota. Larger raccoons can put on more fat and thus survive many more days of fasting during the food-short northern winter. But fasting is not the giant panda's problem. Bamboo is a steady food source, but its low nutritional value comes with its own constraints. Because giant pandas live on a very precipitous edge energetically, they have to conserve all the energy they can muster.

So how did the tropical-living giant panda differ from today's cold-adapted giant panda? The giant pandas we know are superbly insulated against damp cold. For many other northern mammals, a seasonal change in thermal conductance (the ease with which heat is passively transferred from the body through its tissue and pelt) is partly achieved via cyclic changes in the insulating quality of their pelt. In spring and summer they molt, and then regrow their fur in the late summer and fall when food is abundant. The tropical-living crab-eating raccoon does not have a seasonal molt and its pelt has a high thermal conductance. An increase of thermal conductance facilitates passive heat loss. It seems reasonable to assume that a tropical-living giant panda was warm-adapted, more like a tropical-living procyonid.

We believe that, not so long ago, at least two morphs of bamboo-dependent giant pandas existed: those that could live in hot environments (let's call them tropical lowland giant pandas), and those that lived in cold ones, the mountain giant pandas, much like the northern raccoon and the crab-eating raccoon. The northern raccoon diversified its diet and raised its basic metabolic rate to cope with cooler climates. The giant panda could not, because it is ratcheted into a set of adaptations for eating and processing bamboo. The giant panda could, however, tighten up its pelt to prevent heat loss, and that apparently is just what it has done. It does not have a seasonal molt; seasonal molts take a lot of energy, but in species in which they occur, abundant seasonal resources offset the energy they require. The giant panda lives on a steady diet that keeps it always on the nutritional edge. Its molt is fairly continuous, taking a steady but small energy toll.

We watched Mei Xiang hanging raccoon-like over a branch not long ago, and we thought about the raccoon we saw in a similar pose twenty years before. Northern raccoons and giant pandas face a similar challenge: how to keep cool while saving energy. This giant panda has her pelt rigged against her; she is just too well insulated to live full-time in a hot, humid environment. So, giant pandas not only have been ratcheted into a diet of bamboo, but they also have been ratcheted into a life of eating the bamboo that lives in the cool, humid mountains of central China. We have lost the warm-adapted tropical lowland form of the giant panda. A mountain form remains. This limits our conservation options.

4

GIANT GRASS-EATING BEARS

Veils of fog swirl, separate, lift, giving brief glimpses of the steep V-sides of the valley cloaked in a thick, dark-green forest. The forest pinches in on the terrace above the Bei Lu River, where the headquarters of the Tangjiahe Giant Panda Reserve is perched. We see little bamboo in the understory of this evergreen and deciduous broadleaf forest here at 4,600 feet. The bamboo species giant pandas need grow at higher elevations,

Previous pages: Clouds rising against slopes that tower over the western edge of the Sichuan Basin drop their moisture, creating an ideal climate for bamboo.

The panda's thick, slightly oily coat keeps water away from its skin, an important feature in its cool, humid home.

Right: A panda, temporarily full, takes time to rest and digest before continuing the labor of eating up to forty pounds of fibrous bamboo in a day.

around 6,600 feet, and the giant pandas spend much of their time above 8,500 feet. We could have reached the lower edge of giant panda habitat by walking six miles farther up the valley where these bamboo species do grow under a tree canopy. The cool, drizzling rain that had fallen all the night before had everything dripping and clammy, even on this mid-May day: typical weather for giant pandas. The giant panda's short, thick, springy coat keeps it warm and the coat's slight oiliness prevents water from penetrating, an adaptation to the cool, moist climate of these mountains.

When John visited here in 1981, Chinese forestry officials had just phased out a logging commune. The reserve headquarters we visited twenty years later had been the site of the sawmill. Now, the old logging commune building is a service area. There are new staff quarters, reserve offices, and a guest house, and a small museum to introduce visitors to the natural history of the reserve. The once extensive logging-road network in the reserve is now unusable, even by four-wheel drive vehicles. Heavy summer rains and frequent floods have washed out the roads in many places; they have not been maintained because the revenue to do so dried up when logging was banned.

This is good for the fifty or sixty giant pandas living here. But even so, as remote as the giant panda's habitat is in the upper reaches of the reserve, a steady flow of people comes into it. They walk over the high ridges from adjacent Pingwu County to the east or over the main spine of the Min Shan from Gansu Province. They come to gather medicinal plants or snare tufted deer, muntjac, musk deer, and takin. Patrolling these remote areas is demanding, but the Tangjiahe staff is committed and enthusiastic. Just maybe, there is a future for the giant panda here.

THE REAL SQUEEZE

At various times in their long history, giant pandas ranged over much of eastern China. They are now found in only six mountain ranges and are further divided into about twenty-five small populations in a narrow crescent at the eastern edge of the Tibetan Plateau. These are tremendous mountains with soaring ridges punctuated by deep, narrow valleys. The wildlife and habitats in these mountains were once insulated from human influence by the sheer ruggedness of the landscape, but that ruggedness also increases the region's vulnerability and decreases its resilience to humans' disruptive influence. The once continuous forest in these mountains is now reduced and broken, and the giant pandas are in a real squeeze. The lower slopes and valley bottoms have been logged and taken over by farmers. The tiger that once roamed the valley bottoms is gone. Herdsmen burn the alpine grasslands and the fires enter the higher-altitude forests, eating away at the upper forest margins. So all that remains within the mountains for giant pandas is a narrow belt—at places no more than 3,280 feet in altitude—of forest.

This is the northern edge of the distribution of bamboo for this region of Asia and the edge of the Sino-Japanese floristic region. The narrow crescent where the giant panda is now found is uniquely both exposed and protected at the major juncture of China's two major climatic systems. There are no large mountains in southeast China to deflect monsoon rains, but major mountain barriers to the north and west deflect the Asian continental weather system's extreme temperatures and harsh, drying winds. Moisture-laden clouds are driven northwest by the monsoon into the cul-de-sac formed by the western escarpments, spines off the Tibetan Plateau. The clouds rising against these steep slopes drop their moisture, resulting in a total annual precipitation well above that of surrounding regions. This same juxtaposition of mountains and weather systems appears to have a stabilizing influence on what in some adjacent areas is considerable variation in annual precipitation. The resulting climate in the mountains is cool, more so than would be expected at this latitude, but not extremely so. Rainfall is heaviest in summer, but the area never seems without mist or drizzle, or snow in winter. Humidity is moderate to heavy throughout the year. In essence, the macroclimate in the mountains in which giant pandas live is cool maritime, even though the mountains are far inland.

In mountain environments, forest vegetation is a mosaic determined by macroclimate, soil fertility, drainage, and past use by people. Giant panda habitat may be remote to our Western way of thinking about wild places, but to think of giant panda habitat as pristine, untouched, or without a long history of human use is misleading. Rather, we must understand how giant pandas respond to the environmental template where they live, which includes past human use, and be sure no critical elements are missing that might limit their abundance and potential distribution.

The panda's strong jaw muscles power the broad, flat cheek teeth that grind tough bamboo. Like all carnivorans, pandas possess strong canines.

BAMBOO SPECIALISTS

At Tangjiahe's natural history museum, Lucy Spelman carefully examined two giant panda skulls on display

that superficially look like American black bear skulls. Giant panda dentition is almost, but not quite, typical for a bear, with forty-two teeth including incisors, canines, premolars, and molars. The broad, flat molars and premolars, which giant pandas use to grind their principal food, bamboo, are larger and more robust than those in black bear skulls of comparable size. What also stands out on first glance is the prominent sagittal crest above the braincase and wide zygomatic arches, or cheek bones, that form a nearly perfect circle; in other bears, the cheek bones are more triangular. The sagittal crest and zygomatic arches are the attachment points for powerful jaw muscles that let the teeth do their hard work. Lucy, ever the teacher, turned to our group and said, "Look at the difference in the sizes of the skulls. The massive molars in the larger skull are very worn, as are the back premolars. This was a very large animal, but George Schaller noted that the giant pandas he captured here in Tangjiahe for radio-tracking in the 1980s were smaller than those his research team captured at Wolong."

In the larger skull, the broad, flat, and heavily cusped molars and the posterior premolars were worn down from a long life of crunching and grinding the stalks, or culms, of bamboo. The molars of the smaller animal were lightly worn, suggesting it probably died of factors other than old age. Both were skulls of males found in the reserve, our hosts inform us. Adults within a sex vary in size, as these two skulls show. Size varies between the sexes as well. Females

weigh about 198 pounds on average; males on average are 10 to 20 percent larger than females. Compared to the largest male brown or polar bear, which can weigh more than 1,100 pounds, giant pandas seem diminutive. This is part of their charm and why they are more appealing than other bears to so many people.

At 11 pounds, the red panda, the other bamboo-eater that coexists with the giant panda through much of its mountain habitat, is much smaller. (Apparently, they have been extirpated from Tangjiahe.) Red pandas have a very low rate of metabolism and carefully select and eat mostly the most nutritious bamboo leaves. The giant panda eats the robust stalks as well as leaves and shoots. The larger body of the giant panda may enable it to use lower-quality, more mature, fibrous bamboo parts. Largeness also has advantages in cool climates; it takes relatively less energy relative to body size to keep a larger animal warm than a smaller one, because the larger one has relatively less surface area from which to lose heat.

A large body coupled with massive bamboo-crushing jaws should deter most predators, such as the leopards and dholes that coexist with giant pandas. There is one record of a subadult giant panda being killed by a leopard, and there have been reports of aggressive encounters between dholes and a giant panda. Leopards kill sloth bears in Sri Lanka and other areas of the Indian subcontinent where they co-occur. And tiger biologist Dale Miquelle told us of a Siberian tiger that seemed to specialize on killing subadult brown bears in the Russian Far East. Tigers have been extirpated where giant pandas live, but at one time they may have posed a threat to pandas.

The insightful ethologists Ramona and Desmond Morris speculated that the bold black-and-white markings of a giant panda serve as a warning signal, backed up by dangerous jaws and teeth, to potential predators. Coat color in mammals has three known functions: to signal to members of the same species or to potential predators; to camouflage the animal; and to augment thermal regulation. After seeing giant pandas in a variety of circumstances, Schaller and his co-workers at Wolong found that, except in snow, the giant panda's coat did not camouflage it. They thought its function was to signal other giant pandas.

In contrast, the pure white fur of a polar bear (which covers black skin), seems to be an adaptation to living on sea ice, where the bear needs concealment to hunt seals. Polar bears have black noses and foot pads. Using infrared sensors, scientists measuring heat dissipation in exercising polar bears found that the muzzle, nose, ears, foot pads, and the insides of the thighs are "hot spots" that dissipate extra heat. Equivalent measurements have not been made on giant pandas, but National Zoo staff found, much to their chagrin, that giant pandas' coats insulate so well that an expensive motion-detection system that relied on infrared heat sensors did not work—the amount of heat the animals were losing was too small for the system to detect. As in most mammals, though, coat color and pattern serve multiple functions in ways that are difficult to tease apart. Regardless, giant pandas' bold black-and-white markings are stunning and contribute to their universal appeal.

Left: A zoo panda might eat ornamental grasses, but pandas in the wild stick to the woody grasses called bamboo. Bamboo forms 99 percent of a panda's diet.

Depending on the season, pandas eat stems, leaves, or shoots of bamboo.

Scientists once thought that wild giant pandas would become omnivorous when bamboo was in short supply. One observer reported seeing a giant panda and two cubs feeding on gentians, crocuses, and irises, far from any bamboo forest. There were no detailed field studies yet to judge the credibility of the observation and it became embedded in the thinking about giant panda natural history. But these animals were most likely a brown bear and cubs. Wild giant pandas never rear more than one cub and have never otherwise been seen feeding in the open. So, while wild giant pandas eat a smattering of other items, more than 99 percent of their diet is bamboo.

We know from maintaining giant pandas in zoos that they can subsist on foods other than bamboo, so they could potentially be omnivorous like the black and brown bears. But wild giant pandas are not. Zookeepers were concerned about providing a diet of only bamboo, fearing that giant pandas would not be able to select foods that would provide them with all the micronutrients they need. This led to giant pandas being fed large portions of concentrated food in pellets, a practice that has fallen out of favor. Most zookeepers now maximize the amount of bamboo in the diets of their giant pandas. Conservation scientists are concerned about whether giant pandas living or born in breeding centers and zoos that have been eating mostly concentrated foods can learn how to select and subsist solely on a bamboo diet. This would be essential if these animals are needed for reintroduction, so it seems more sensible to maintain them on bamboo. While reintroduction is not needed now, it is an option conservation scientists are exploring.

Some people compare the giant panda's specialization on bamboo with the marsupial koala's specialization on eucalyptus leaves, but there are significant differences. The koala's physiology is so closely adapted to a diet of eucalyptus leaves that this animal cannot change readily to more conventional foods. But for giant pandas, "Their dependence on bamboo reflects the absence of a large alternative food source, especially in winter, rather than an inability to assimilate other foods," Schaller and his Chinese associates concluded after their intensive study of giant pandas in the Wolong Reserve. This ability to subsist almost exclusively on the stems, branches, and leaves of bamboo makes the giant and red pandas unique among the Carnivora.

Where Asiatic black bears and giant pandas coexist, they feed on different diets: black bears eat herbs, fruits, and nuts; giant pandas eat bamboo. But black bears gain two to three times more energy per pound from their food than pandas do from theirs. So the obvious question is, Why does the giant panda restrict its diet to bamboo? The answer lies in the availability of the various foods. The bears' primary food sources vary significantly in their abundance from one year to the next, and their availability changes during the annual cycle. Bamboo is lower in quality, but it is in constant supply. For example, living on fat reserves built up in seasons of plenty, black and brown bears hibernate to get through long winters; giant pandas instead spend their winter days steadily eating bamboo. Schaller characterized the difference this way: "Bears have remained opportunistic and adaptable, geared to a boom-or-bust economy, evolving a lifestyle that has enabled them to settle and thrive in many different habitats. The panda has become a specialist, it has chosen security over uncertainty."

In essence, giant pandas are big grass-eating bears. Their natural history reflects the costs and benefits of subsisting on large, woody grasses.

LIVING LARGE ON WOODY GRASSES

How the giant panda can survive as a herbivore with the digestive system of a carnivore has been the great enigma for scientists trying to understand its biology. It shouldn't work, but it has. In most of their morphological and life-history traits—such as locomotion, brain size, skull size, and age at sexual maturity—giant pandas are not unusual for carnivores in general and quite similar to bears, biologist John Gittleman concluded. The real biological distinction is in how they manage to live on a diet of bamboo.

If a giant panda comes across a suitable alternative food item, it will eat it. About 1 percent of a giant panda's food in the wild comes from other sources, including an occasional bamboo rat or musk deer fawn, but these opportunities are very rare in the bamboo thickets where they spend their time. In 1999, on the other side of the high ridge above the Tangjiahe Natural History Museum, about 18 miles to the west in Pingwu County, we saw three goats killed by a young female that was probably dispersing from Tangjiahe. But the young female had not eaten the goats and her killing bites lacked the precision of those of a big cat.

The other members of the Carnivora either eat mostly meat, such as the obligate carnivores like the cats, or they are omnivores that eat a mixture of meat, fruits, and seeds—foods rich in protein, lipids, and starches. These carnivores digest these relatively easy-to-digest foods in simple stomachs and short intestines. The intestines are short because long retention times are not necessary to absorb all the nutrients from these foods. On the other hand, herbivores subsisting on leaves, stems, and twigs have to cope with food that is both low in nu-

Bold black and white markings are probably signals to other pandas—and perhaps potential predators—not camouflage.

trients and hard to digest. The intestinal length of a giant panda is five to seven times its body length, while a domestic cat's intestine is about four times as long as its body. In contrast, a horse's is about ten times, and a cow's is about twenty times its body length. Clearly, the giant panda is not nearly so well adapted an herbivore as a horse or a cow, and not much better than a cat that eats only meat.

An herbivore can obtain energy from two primary plant components: the cell contents with their soluble nutrients, and the tough, fibrous cell walls composed of cellulose, hemicellulose, and lignin. Herbivores such as horses and hares have modified hindguts that store, ferment, and absorb nutrients. Cattle, deer, and other ruminants have a specialized foregut that is effectively a vat where food is fermented before passing into the stomach for further digestion by enzymes. In both of these adaptations, the herbivore has a symbiotic relationship with certain bacteria and protozoans that can digest cellulose and hemicellulose—the primary constituents in the walls of plant cells—by fermentation and synthesize amino acids using nitrogen, which the herbivore then absorbs. The microbes cannot digest lignin, which passes through as waste.

A carnivore-turned-herbivore, such as a giant panda, which has a carnivore's simple stomach and relatively short intestine, does not have the physical and physiological adaptations to process a bulky herbivorous diet. Lacking the ability to digest the cellulose in the cell wall, the giant panda depends on the soluble nutrients in the contents of the plant cell rather than those in the cell wall. The panda is able to obtain nutrients from only one cell wall component—hemicellulose, a complex polysaccharide—and this constitutes 23 to 35 percent of bamboo leaves and stems. Acids and alkalines in the giant panda's stomach and intestines break down some of the complex sugars, enabling microbes in the lower gut to digest a small percentage through fermentation during its brief passage.

Unlike most grasses, which vary throughout the year in their nutrient content, bamboo is relatively constant in the amount and quality of nutrients, or at least it is in the way that giant pandas select parts to subsist on. The Wolong Giant Panda Project scientists found they could divide wild giant pandas' year into three seasons based on bamboo selection. From April to June, pandas ate arrow bamboo stems and new umbrella bamboo shoots; from July to October, they ate mostly leaves; and from November to March, they ate old stems of arrow bamboo and leaves. They found that giant pandas obtain ample protein from their food, so why giant pandas eat so much is explained by their need to obtain sufficient calories. A 220-pound giant panda needs about 2,200 calories (technically kilocalories) per day just to live in a resting state; this

Pandas spend more than twelve hours a day active, with foraging and feeding the dominant activities. They may travel as little as the equivalent of five or six city blocks in a day.

is termed its basal metabolism. Energy for growth and reproduction and for other activities raises this total requirement to between 3,500 and 4,000 calories per day. For comparison, a person the giant panda's size, engaged in moderate activity, requires about 3,000 calories per day; the average minimum consumption in the United States is 3,700 calories. Average daily bamboo intake was estimated to provide between 4,300 and 5,500 calories per day, considered a low margin of safety by Schaller and his associates. These estimates are essential to understanding the giant panda's natural history and energetics, which can be advanced by studying pandas living in naturalistic zoo habitats. Because wild giant pandas travel so little every day, the zoo setting fairly well reflects the wild giant panda's energy needs.

EATING BAMBOO

The giant panda's head and forepaws are specialized for feeding on bamboo. The broad head with a short muzzle is the anatomical consequence of the massive jaw muscles needed to masticate the woody stems of bamboo. The giant panda's sixth digit, the "thumb" on its forepaw, is a greatly enlarged wrist bone, the radial sesamoid. This digit is capable of independent movement and, used in conjunction with the first digit, enables the giant panda to handle bamboo stalks with great precision.

National Zoo visitors watch with avid curiosity while the giant pandas eat their bamboo. Bears eating bamboo, like the gigantic elephants across the way, seem surreal.

Giant pandas possess an acute sense of smell, and they may sniff at a bamboo stem as they bring it to their mouth. Using the forepaws, a panda detaches leaves from stems and then pushes the leaves into its mouth. In the wild, a panda usually sits, then hooks a bamboo stem with the curved claws of its forepaw, holds it, and, bending it sideways, bites it off close to the base. Sitting Buddha-like or half lying on its back, the animal pushes the stem at right angles into its mouth. It takes rapid bites and, while biting, it jerks the forepaw holding the stem up and down, and nods its head up and down, to sever pieces of stem. It gives each mouthful a few chews, then swallows. In this way, it eats most of a stem. Year-old stems, which have the best balance of amino acids, provide the bulk of the panda's diet. Schaller and his associates found that in Wolong, a giant panda spends fourteen or so hours a day searching for, selecting, and eating the 22 to 40 pounds of bamboo it must have each day to survive.

Giant pandas spend about 60 percent of their time active, but they move each day only about 550 yards, equivalent to walking five or six city blocks. During that time, they are mostly feeding; their other activities, including drinking, traveling, scent marking, and grooming, take only a small portion of their day. They can be active anytime, day or night, but the scientists working in Wolong found that they are most active between 4 and 6 A.M. and between 4 and 7 P.M. The activity times of the mostly solitary giant pandas are not synchronized with those of their neighbors. The remainder of the day, giant pandas are inactive to conserve energy. They take long rests that usually last two to four hours, but sometimes as long as six hours.

What limits a giant panda's activity and determines the length of its rests: the time it takes food to pass through the gut, or the quantity of food it can process? Giant pandas are limited by the quantity of food they can process. It takes about six to eight hours for food to pass through the digestive tract, from the time a piece of bamboo goes in the animal's mouth until it comes out the other end. They can increase intake with highly digestible foods, such as bamboo shoots, but their feeding strategy through most of the year is slow and steady, focused on abundant but predictable bamboo leaves and stems. To survive on a diet of bamboo, pandas have to keep their digestive tracts full and are constrained by the maximal volume they can process through their digestive system. They digest only about 17 percent of the plant dry matter they ingest, a very low amount for a herbivore. During the course of a day, they obtain from 8½ to 22½ ounces of protein from their bamboo diet. Schaller and his associates concluded that "the giant panda consumes large quantities of bamboo more to obtain calories in the form of soluble carbohydrates than to fulfill protein requirements." Pandas do require water. The moisture they excrete in their feces is far more than they take in through the bamboo they eat. In summer, their mountain habitat is often dripping after frequent rains and they do not have to drink, obtaining their water from wet bamboo leaves. They do need to drink at least once a day in winter, but they do not appear to eat snow.

Giant panda bedding sites are so characteristic that scientists in the mid-1980s estimated panda numbers by counting the bedding sites. Pandas

Left: Scientists suspect giant pandas must have low metabolic rates to subsist on nutrient-poor bamboo.

Bamboo-eating red pandas are known to have low metabolic rates, comparable to those of sluggish reptiles.

do not make nests, but often rest at the base of a tree, and their droppings are found around the perimeter of their beds. After a long rest, between ten and fifty droppings may accumulate. They reuse their good rest sites. The giant panda survey teams used the figure of forty-five bed sites per month for each giant panda to calculate the number of pandas living in an area. Giant pandas climb trees to escape other pandas or while courting, but they also occasionally rest and sun in trees.

Watching giant pandas sleep at the National Zoo, boring as it sounds, seems to delight visitors. The pandas rest on their bellies, all sprawled out, or on their backs with their legs tucked in, or partly on their sides, or with one hind leg in the air, or with their forepaws covering their eyes or muzzle. Visitors seem to relate to their cavalier sleeping styles. Sleeping pandas seem completely relaxed but often change position. Some of the changes may be adjustments to conserve or dissipate body heat. But while giant pandas seem the ultimate slackers, they are actually on a very tight schedule. To survive, they must balance how much food they eat, its digestibility and nutritional value, and the time it takes to digest that food—and this is what a giant panda in repose is doing.

So giant pandas have opted for one of the strategies used by herbivores

Towering, clump-forming bamboos, which thrive in the sun and usually occur at low elevations, are not much eaten by pandas.

living on low-nutrient plant food: they consume the most nutritious parts of the bamboo—leaves and year-old stems—and they consume prodigious numbers of new bamboo shoots in season. (The latter creates a direct conflict with the people who enter their forest domain to do the same.) Some mammals adapted to survive on nutrient-poor plant foods, such as the leaf-eating koalas and three-toed sloths, have managed to do so by lowering their basal metabolic rate and restricting their physical activity. Brian McNab of the University of Florida found that the red panda restricts its physical activity and has a basal metabolic rate about one-third that of the similarly sized rabbit-eating bobcat. As McNab phrased it, "The paradox is that the red panda maintains a high, regulated core temperature at low ambient temperatures while having a metabolism similar to that found in reptiles." Heat production can be reduced only if there is diminished heat loss, and the red pandas apparently accomplish this with thick fur and a reduction in skin temperature. Do giant pandas have similar physiological adaptations? We don't know, but it is tempting to speculate that they might, given their low activity levels and the high insulation value of their fur coats.

Our conservation strategies for the giant panda in the wild have to be based on understanding its needs. We must come to understand this relationship between giant pandas and bamboo—its constraints as well as advantages—if we are to sustain wild giant pandas.

BAMBOO BASICS

Bamboo is so ubiquitous that the supply seems limitless. How can the future existence of the giant panda be limited by something we think of as so abundant?

Bamboos are evergreen, woody, branching grasses. China and bamboo seem synonymous in our Western minds, but bamboos live throughout the humid tropical and temperate regions of the Western Hemisphere, Africa, and Asia. They cannot live where there are extended periods of low humidity or extreme cold. Worldwide, there are an estimated 2,000 species (although the number has been disputed for years), and more than 100 species thrive in Sichuan alone.

Unlike typical grasses that flower and produce seed annually, most woody bamboo species flower gregariously. This means that most, but not all, members of a particular species will flower, produce seeds, and then die in synchrony throughout their range. During the 1983 mass flowering of bamboo at Wolong, 10 percent of the arrow bamboo survived; giant pandas survived by switching to other bamboo species that were not flowering. Bamboos flower and die at intervals that range, depending on the species, from 3 to 120 years. What triggers flowering remains a mystery, but many suspect a correlation with drought. J. J. N. Campbell and Z. S. Qin suggest that because the adult plants would suffer from lack of water anyway, they choose that time to reproduce and die.

After they flower, most colonies die, and new ones are established from the seed that has been produced. Wind-pollinated, bamboos must be cross-pollinated to produce fertile seeds. It may take years for new stems grown from seed to be tall enough to serve as food for giant pandas, and a stem may live for fifteen years. To spread between flowerings, bamboos reproduce vegetatively via rhizomes, root-like stems. The rhizomes develop underground and, once a year, the bamboo sends up shoots that emerge from sheaths to grow into culms. In the temperate zone, shoots are a spring or summer event, and in many species these shoots grow incredibly fast—the record is a Japanese bamboo culm that grew almost four feet in a day! In contrast, bamboos grown from seed may take five to twenty years to reach full size and maturity. In Sichuan, people regard bamboo flowering as inauspicious: it predicts privation for years to come.

Why bamboos flower and die gregariously may have more to do with seeds than flowers. When bamboos finally get around to it, they may set massive crops of seeds. A 40-square-yard clump of one Indian species can produce more than 300 pounds of seeds—at between eight hundred and one thousand seeds per ounce! One oft-told tale is of a surveyor who had to quit working when the pear-sized seeds of another Indian bamboo were falling so thickly, over a 6,000-square-mile area, that they were breaking his instruments. Most bamboo seeds are small, however, and serve as a good substitute for rice. Having put prodigious amounts of energy into seeds, the bamboos are spent, and die.

Ecologist Dan Janzen proposed that massive seeding in synchrony may have evolved in response to the multitudes of seed predators, from rodents and birds to rhinos and elephants and people, that find bamboo seeds appealing. A bamboo producing only a few seeds at a time risks having all of those seeds eaten. But huge seed crops swamp predators—they simply can't eat them all—so many seeds survive to grow into plants. A bamboo that produces seeds when all the other bamboos are producing seeds will see more of its seeds survive than a bamboo that drops seeds alone. Another explanation is that flowering all at once increases the chances of successful reproduction in this wind-pollinated plant. Yet another idea is that few seeds would survive among the dense culms of adults, so many seeds—and the parents' death—are necessary for the successful reproduction.

The bamboos generally fall into two types: sun-loving and shade-loving. The sun-loving bamboo species grow in clumps in the open and can exceed 33 feet in height. Shade-loving bamboos influence forest succession in panda habitat because the growth of their rhizomes suppresses the growth of trees, particularly conifers. These bamboos also colonize areas in the forest with a history of disturbance. In most wet tropical forests, encountering a stand of bamboo immediately indicates the area was once a farm or homestead.

Bamboos are vulnerable to predation by herbivores at all times of the year. Like all grasses, they have no toxic secondary compounds as defense against predators in their stems, but are protected in part by sharp silica spicules, tough woody stems, and coarse leaves. Silica and tough woody stems eventually take their toll on a giant panda's teeth, as we saw in

Left: A few stems of bamboo mixed with profuse ferns at the lower edge of panda habitat indicate that local people sometimes burn this area to clear land for grazing animals.

Bamboo thickets growing under conifers on mountain terraces and basins are ideal panda habitat.

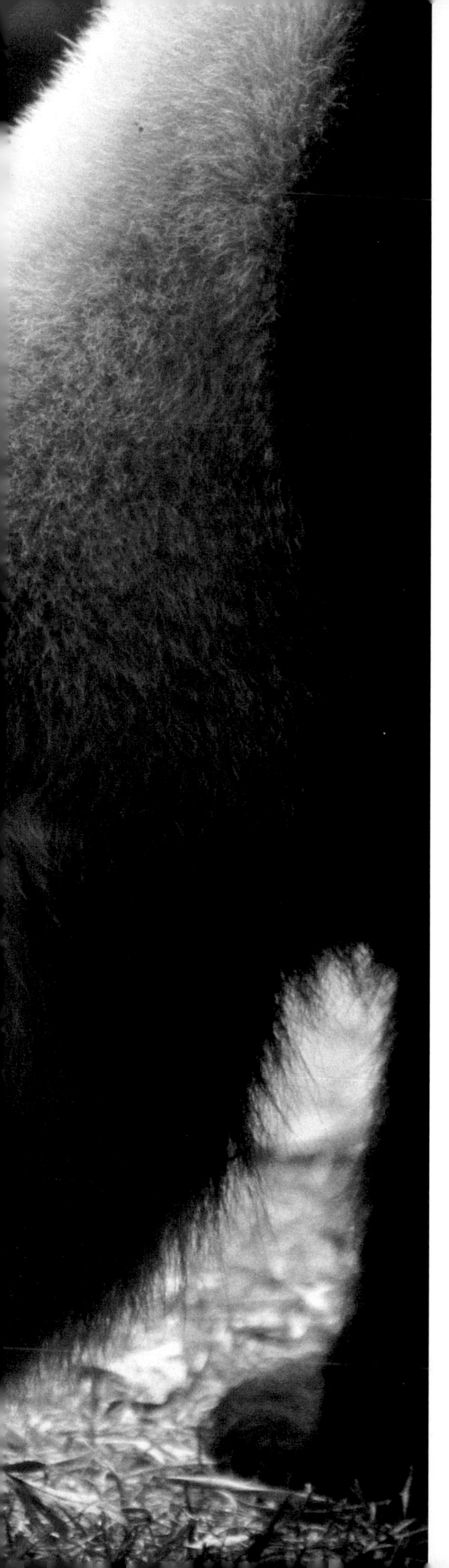

the skull of the old male at Tangjiahe.

Most bamboo shoots have some amount of cyanide, although it is a form less poisonous than that of murder-mystery fame. This raises the question of how giant pandas and other shoot-eaters manage to detoxify them. We couldn't find any information on the amount of prussic acid in the shoots of the bamboo species giant pandas eat, but some other species have poisonously large amounts. The golden bamboo lemurs of Madagascar specialize on the shoots of a giant bamboo and ingest enough cyanide to kill other mammals; neither of two other species of bamboo lemur eat these shoots. How the golden ones cope is unknown.

Mountain gorillas also savor cyanide-laced bamboo shoots, which may form more than 90 percent of their diet when the shoots are available during the dry season; they also have a lot of diarrhea when they're eating shoots, and this coincides with observations of mountain gorillas eating soil, a practice called geophagy. Analysis of the soil revealed properties similar to the ingredients of the anti-diarrhea medicine Kaopectate. Schaller reported finding claylike soil in the droppings of one wild giant panda and observed a panda in a Chinese zoo eating similar soil in an outdoor enclosure.

Because giant pandas eat so many bamboo shoots and stems, they affect the bamboo stands on which they feed. In Wolong, Schaller and his associates found that the new growth in bamboo shoots each year in arrow bamboo stands was enough to support about five giant pandas per square mile. In umbrella bamboo stands, insects and giant pandas destroyed one-third to one-half of the annual shoot production. Giant pandas forage along the edge of thickets and on thick shoots. In doing so, they slowed the rate at which bamboo thickets spread and favored the survival of spindly stems.

Before we can examine how giant pandas use their mountain habitats, we need to understand the dynamics of bamboos growing in these mountain forests. For Western biologists and conservationists, this is an unfamiliar and puzzling dynamic ecological system, and Alan Taylor and Qin Zisheng spent several years working it out in giant panda habitat in the Wolong Reserve. Most of the bamboos preferred by giant pandas are shade-loving species. For example, in the northern areas of their range, they prefer to forage on arrow bamboo growing beneath a forest canopy, even though stem densities are greater in open, clear-cut areas.

Bamboo can grow in stands of a million stems in a square mile area. The thick stands of bamboo growing in clear-cut areas reduce natural tree generation, especially of conifers. Even under an old-growth conifer forest, mature bamboo reduces tree regeneration. Spruce-fir forests that had been clear-cut in the 1930s, as well as recent clear-cuts with arrow bamboo

Pandas generally forage alone, in their own daily rhythms, and come together only briefly to mate in the spring.

understories, are now solely composed of various species of birch as the canopy tree species, suggesting that clear-cutting leads to development of a persistent hardwood forest dominated by birch. Birch readily colonize clear-cuts because their seeds are winged and wind-born, their saplings grow faster, and timber harvesters leave few if any coniferous seed trees.

Loss of the dominant conifers is detrimental to giant pandas because they use hollow old conifers as maternity dens. They need trees that are 3 feet in diameter; a tree that size may be 200 years old. And they like to feed on the more robust bamboo stalks that grow under conifers, rather than the smaller ones that grow under a hardwood canopy or in clear-cuts. Loss of conifers is also a problem because it reduces bamboo seedling survival. Seed establishment after arrow bamboo flowered in the early 1980s, as it does every forty-five years, was much lower in areas that did not have a forest canopy.

Many people predicted that the mass flowering of arrow bamboo in the northern Min Shan in the 1970s would be catastrophic for giant pandas. The Chinese Ministry of Forestry put a program in place to rescue starving giant pandas and reported finding 138 dead. And, indeed, where there was only a single species of bamboo and it mass flowered, up to 80 percent of the giant pandas reportedly died. But in areas where there were two or more bamboo species and one did not flower, there was little mortality. Schaller emphasized, ". . . forest cover is needed if bamboo is to recover after a die-off, and each area should have at least two bamboo species upon which pandas can subsist." Denning sites may also be limiting.

The real challenge facing the giant panda is loss of habitat, especially those limited areas where at least two bamboo species grow. A comparison of surveys done in 1975 to 1976 and from 1985 to 1988 revealed that the habitat occupied by giant pandas had been reduced during that time by about half, to 5,000 square miles. It was further reduced through the late 1980s and 1990s. This largely accounts for the giant panda's plight.

GIANT PANDA BEHAVIOR

In the 1980s and into the early 1990s, two research projects provided nearly all we know about wild giant panda behavior and ecology. The first project was led by Schaller and his associates Hu Jinchu, Pan Wenshi, and Zhu Jing in the Wolong Reserve in the Qionglai Shan and in the Tangjiahe Reserve in the Min Shan, both in Sichuan Province. The second was led by Pan Wenshi, Lu Zhi, and their students on the southern slopes of the Qinling Shan south of Xi'an in Shaanxi Province. Before these groundbreaking studies, our knowledge of giant panda ecology and behavior was limited to narratives written by the panda hunters as they searched for giant pandas to shoot, including their vivid (if disturbing) accounts of how the pandas they shot died.

Seeking out giant pandas in their mountain bamboo thickets is extremely hard work, as the early hunters repeatedly remarked upon. Very few people will ever see a giant panda in the wild, no matter how hard they try. Chris Catton noted that, through 1990, fewer than fifty Westerners had ever seen a panda in the wild. Satellites passing over, gathering information on forest types and changes, are essential tools to understanding the rate and quality of changes in these mountains, but satellite images cannot tell us how giant pandas actually respond to environmental change or to subtle differences in vegetation, slope, exposure, and history of land use by people. Because pandas are retiring and elusive, the Chinese government allowed the field biologists working in Wolong, Tangjiahe, and Qinling to attach small radio-transmitter collars to a few selected individuals. They captured the animals using box-traps baited with cooked mutton, of all things, but it worked. From these few radio-tagged animals, we now have an understanding of how the reclusive pandas respond to their environmental template.

We also have the observations made on giant pandas living in zoos, but the very nature of the zoo environment and the limited number of animals in any one place restrict how a giant panda can respond. In spite of these difficulties, behaviorist Devra Kleiman produced a remarkable description of the behavior of Hsing-Hsing and Ling-Ling during the first decade of their lives at the National Zoo.

In his landmark book, *The Mammalian Radiations,* John Eisenberg, former National Zoo Resident Scien-

Pandas communicate with one another without meeting by leaving scent marks on tree trunks and other objects. Scents may indicate a panda's sex and status.

tist and Assistant Director for Animal Programs, demonstrated how useful it is to think of a mammal's life in terms of five behavioral systems: foraging, mating, rearing, dispersing, and refuging—where it seeks shelter to rest and to escape predators, or severe weather, for example. Our descriptions of giant panda behavior are organized around these systems.

In terms of the giant panda's foraging system, we need to ask how large an area a panda uses. Is it shared with other giant pandas, or are parts or all of it used exclusively or defended as in a territory? Giant pandas live in very thick cover and are retiring. How do they communicate to come together or to avoid each other?

What these scientists found after years of effort was that the amount of forest that a panda uses is dependent on its quality, as the giant panda perceives it. In Wolong, female home ranges average about 3 square miles; adult male home ranges varied from 2½ to 4 square miles. The pandas preferred areas with at least 66 percent forest-canopy cover over the bamboo they were feeding on; they avoided former clear-cut areas, even though the bamboo growing there was very thick. They preferred gently sloping terrain with gradients of less than 20 percent to steeper hillsides. In these mountains, gently sloping terraces are not abundant; very steep hillsides are.

Black eye patches enhance the staring threat display pandas use to intimidate each other. Conversely, an intimidated panda may cover its eyes with its paws.

But not all giant pandas use their environment in the same way. A male radio-tracked in Tangjiahe had a total range of almost 9 square miles because he moved far up into the mountains in summer but spent the winter in a single 250-acre area in the valley.

Pan Wenshi and his team found a different pattern of habitat use in the Qinling Shan. Two types of habitat were used by giant pandas, depending on the season. In winter and spring, giant pandas lived mainly in the conifer-broadleaf vegetation zone between 4,400 and 6,600 feet, where they were eating mainly bashan muzhu bamboo *(Bashania fargesii)*. They lived in this habitat for more than 260 days a year on average, in home ranges that were usually less than 4 square miles. In summer, they moved up to between 7,900 and 10,200 feet, where their principal food was songhu bamboo *(Fargesia spathacea)*. Their home ranges in summer were usually about 250 acres in size, although one adult male used nearly 5 square miles. Over a full year, pandas in the Qinling Shan occupied an average of a little less than 4 square miles, with the exception of the one male that used an 11½-square-mile area. For comparison, a tigress living in the very richest habitats in terms of numbers of deer and wild pigs in Nepal needs an area of about 8 square miles to support her cubs and herself. In prey-poor areas such as the Russian Far East, a tigress needs about 200 square miles to live and rear cubs.

Giant pandas move about with a fast or slow diagonal walk. The walk is bear-like but the stride is longer, with much more lateral rotation of the shoulders and hips; this gives them a pigeon-toed look. (In 1938, the Bronx Zoo rejected Su Lin, the first panda ever brought to the United States, partly because staff there believed her pigeon-toed appearance was a deformity; the asking price was also daunting.) Pandas can gallop, but do so only rarely; they can stand erect like other bears, but do so very infrequently. Bears are plantigrade in that their heels touch the ground. The panda's forefeet are plantigrade, but the heels of the hind feet do not touch.

Giant pandas live solitary lives in their bamboo thickets, except during courtship and when females have young. The home areas of both sexes overlap, although it is not clear if they defend their core areas in some circumstances, or if, under some ecological conditions, there is site-specific dominance among both adult males and females. How do giant pandas sort this all out? Mostly through scent marking with urine and anal-genital gland secretions. As Kleiman first described, a panda urinates and rubs its anal-genital gland on surfaces in four ways: it squats and lowers its hindquarters on a horizontal surface; it backs up to a vertical surface; it cocks one leg and backs up to a vertical surface; and it does a handstand. Males handstand and mark by urinating with only incidental anal-genital contact; females do not do handstands. Both males and females have large anal-genital glands, kept tucked away under their broad, short tails

During mating, a male may bite the female's nape; after mating, the female may turn and try to bite the male.

when not in use. These glands secrete a waxy, sticky substance (with a volatile component) that adheres well to surfaces. Within a few minutes of entering his new habitat at the National Zoo, Tian Tian was sniffing at the old marking sites on a rock wall that had not been marked since Hsing died more than a year earlier.

The mode of chemical communication in giant pandas was further elaborated through creative experiments at the China Research and Conservation Center for the Giant Panda at the Wolong Reserve by Donald Lindburg and Ronald Swaisgood of the Zoological Society of San Diego, and Zhou Xiaoping from the Center. Giant pandas body rub, stroking substrates with the head, nape, shoulders, and dorsal surfaces. They scrape the ground with a backward motion of the hind paws. Males scent mark year around; females, most frequently during the mating season. Urine is also used to mark, and females urinate most often during estrus (heat). Giant pandas use communal scent-marking stations where several neighboring animals deposit and investigate scents. Giant pandas can discriminate the scent marks of different individuals. And males can determine the reproductive state of a female through her scents. While most scientists have assumed that these chemical cues are a primary communication channel in

mammals, they have been slow to actually establish this as a fact.

As other solitary mammals do, giant pandas leave olfactory and visual signals through their home ranges. They create scent trees by stripping off bark, biting, and clawing; they paw the ground and rub and roll. Giant pandas may use the ability to distinguish individual differences in scent to mediate interactions among them. These include determining the sex and reproductive condition of an individual and detecting a stranger in their home ranges. A female uses scents to avoid strange males. Male scents may facilitate a female's readiness to mate or allow a female to form mate preferences. Other scents allow a parent to recognize its offspring, and vice versa, and all pandas to recognize kin and avoid inbreeding. The Wolong Breeding Center studies emphasized that "Females' obvious interest in male scent marks, coupled with their dramatic increase in marking behavior during estrus, bolsters the hypothesis that chemical communication in female pandas is mainly important for attracting and/or choosing a mate."

How giant pandas extract information from scent marks or urine remains to be determined. They lick, sniff, and/or display flehmen behavior when they investigate scent marks. In flehmen, the giant panda deeply inhales and curls back its upper lip. This engages its vomeronasal organ through tiny openings just behind the upper incisors in the roof of the mouth. We sensory-deprived primates don't have this organ, though most other mammals do, so we can only guess at the full richness and quality of the information being transferred this way.

Body rubbing and handstand urine marking are used mostly by males. The handstand posture appears to be associated with a male's age and dominance status. "My mark is high, so I must be old and big," seems to be the message. Body rubbing, on the other hand, appears to be a means to make a giant panda's own body odor more like the odor of its environment. Males body rubbed on areas with noticeable concentrations of scent marks left by other males, so they may be trying to smell like a territory owner and bluff him. Females investigated and displayed flehmen to female odors as much as males did.

All giant pandas investigate scent marks they find, but only males and estrous females spend a lot of time scent marking. Male giant pandas actually seek out female core areas outside the breeding season to deposit scent. Foot scraping appears to be associated with aggression between males, and may be a visual as well as a chemical cue.

These scientists concluded that odors in giant pandas boost sexual motivation and cause a shift to using vocalizations to attract a mate. Giant pandas vocalize infrequently, except during mating, but they do have a diverse repertoire of calls. Gustav Peters and Devra Kleiman at the National Zoo first described the repertoire, and it was confirmed, clarified, and expanded in the field by the Wolong Giant Panda Project. Giant pandas huff, snort, and chomp. These sounds reflect the animal's emotional state, grading from apprehensive to mildly threatening. Giant pandas honk when they are mildly distressed and squeal when threatened or attacked. Both calls signal a lack of aggressive intent at different levels of intensity. Their bleat call is the only one that seems to signal friendly overtures. The loud moan, bark, and chirp calls are probably longer-distance advertisements and are associated with the mating season.

At the Wolong Giant Panda Breeding Center, adult males utter a bleat call to promote contact and appease females. Males don't bleat to other males or when presented with the urine of other males, but they bleat to the urine of both estrous and anestrous females. Once a male has detected a female's scent, it then uses vocalizations to initiate contact. The chirping of females is motivated by sexual condition, and females chirp more when exposed to male odors than to female odors.

These calls promote social contact. A growl, on the other hand, is an aggressive threat. Male giant pandas roar during disputes over a female; females roar at persistently courting males. Roars signal an aggressive threat at the highest level. During the first month of life, infants squawk when they want attention, a sound hard to ignore.

The familiar tail movements,

A newborn panda weighs only about three and one half ounces, just 0.12 percent of its mother's weight.

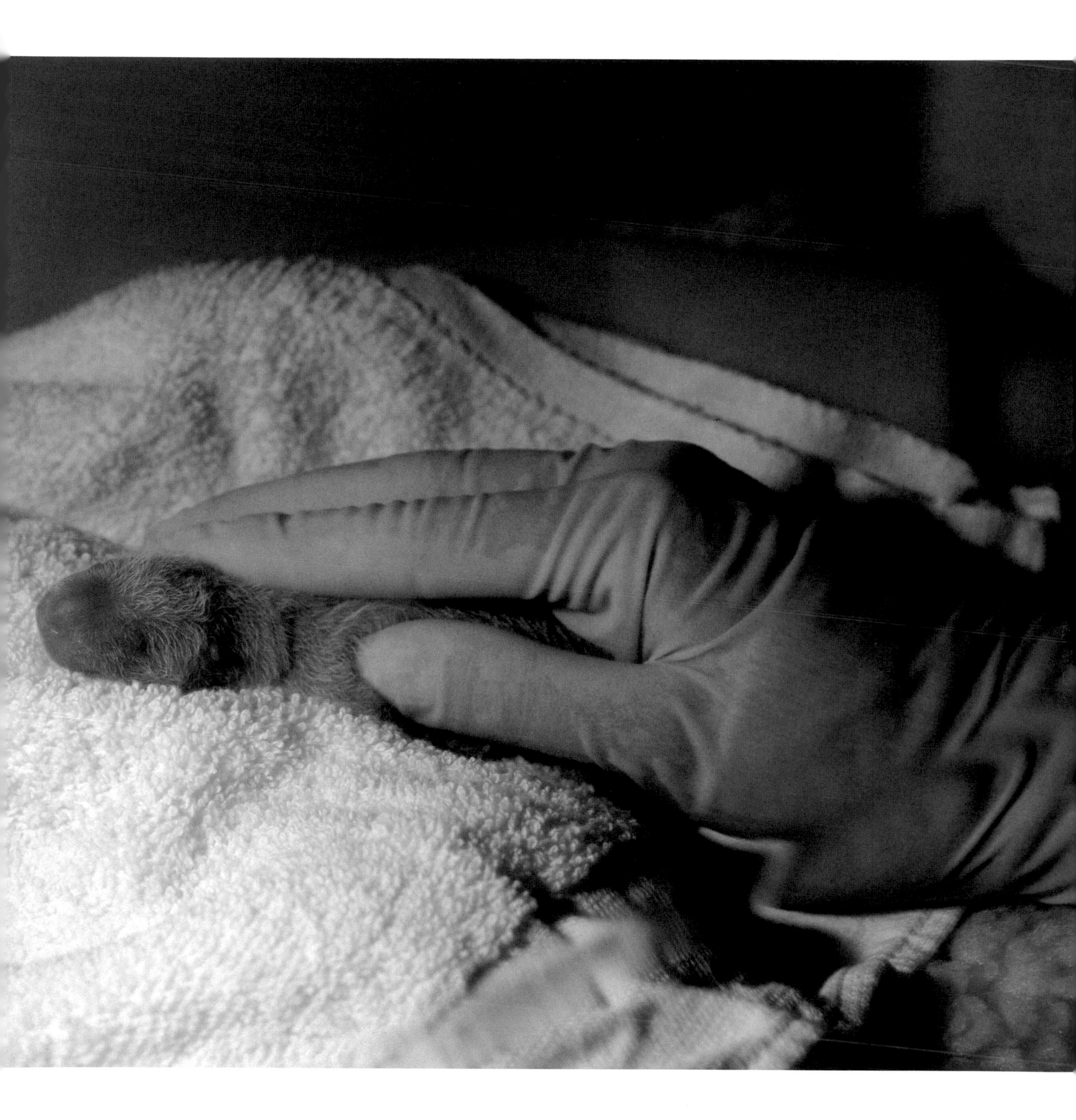

In the wild, a female panda gives birth in the fall, within a large hollow conifer or a cave. She stays with her cub continuously for its first two weeks of life.

Right: A panda mother lavishes care on her cub for its first one hundred days, making it impossible for a female to raise twins.

gestures, body postures, and facial expressions that cats and dogs use to express their moods are shared by many carnivores. When one giant panda threatens another, it lowers its neck or bobs its head up and down. Opponents may circle each other. They swat, rear up on their hind legs, lunge, and grab an opponent. As a sign of appeasement, a giant panda hunches up and tucks its muzzle between its forelegs. When intimidated, it may cover the black eye patches with the forepaws, as does a raccoon. The black eye patches, which enlarge the apparent size of the eye by a factor of ten, may emphasize the staring threat display. The panda does not have mobile ears and lips or well-developed facial expressions, although these are important in close communication in other carnivores.

A giant panda may roll over on its side to indicate a nonaggressive intent. It somersaults and rolls as an invitation to play. Individuals who are subordinate may be forced into rolls and somersaults, or lie on their side and back, usually squealing. Given their basic bodies, they have limited options for expressing themselves. Scientists have exhaustively compared giant panda behavior to that of the other bears, members of the raccoon family, and to the red panda to divine the relationship of the giant panda to other members of the Carnivora, but no clear patterns have emerged.

PANDEMONIUM

In the late 1970s and through the 1980s, spring at the National Zoo, and in Washington, D.C., was heralded by Ling-Ling's and Hsing-Hsing's attempts to mate and, finally, by accounts of their successful matings. This annual spring event was widely and continually covered not just in the local press, but also internationally. Of all the news coming out of Washington, John was able to follow giant panda mating season at the National Zoo in the *Straits Times,* a newspaper printed in Singapore, when he was working to establish national parks in Indonesia in the late 1970s.

Why all the fuss? There is a general belief among the public that sex is a problem for giant pandas. This, of course, is not so for wild giant pandas, but the impression comes from media coverage of the amorous adventures of the few pandas that zookeepers have tried to breed, based on very limited knowledge of how giant pandas actually accomplish this for themselves in the wild. Giant panda females are in heat for only a few days each spring between mid-March and mid-May. A female who does not conceive in the spring may display a weak fall estrus. However, adult male testis size and ejaculate volume decrease through the summer and fall, so whether mating occurs, or whether the males are fertile, is unknown.

For a week or two before she is ready to mate, as her estrogen levels are rising, the female becomes restless, loses her appetite, scent marks frequently, bleats often, and occasionally chirps. She rubs her reddening and swelling vulva against objects and with her paw. Estrogen levels peak

just before her maximum sexual receptivity.

Kleiman described breeding behavior in detail. As a female approaches peak receptivity, she becomes less assertive and allows a male to mount. When copulating, she stands quietly, lifting her tail and depressing her back into a posture called lordosis. The male is responsible for initiating and terminating contact. The female presents her rump while backing up to him or reaches for him gently with her forepaws as she rolls and squirms on her back. Giant panda males have comparatively short penises; their baculum (a bone in the penis) is short, almost pudgy, unlike the long, thin bacula of other bears. Giant pandas do not have a true scrotum; this has confused some zookeepers as to the sex of the animals in their care.

During copulation, the male squats or stands with his forepaws propped on the female's back. He mounts often, but briefly, unlike other bears, which have prolonged mountings. The male sometimes tries to grasp the female's nape with his teeth. After copulation, the female may turn to bite him and run away from him. Males do not tend females after copulation. As far as we know, females are spontaneous ovulators; that is, ovulation is not induced through repeated copulations and stimulation, as it is in cats, for example. Like other bears, giant pandas exhibit delayed implantation. The fertilized egg in the female's uterus develops only to the few cells of the blastocyte stage and then floats free for some time before implanting on the uterine wall.

All this seems reasonably typical of mating in many mammal species, but in zoos and breeding centers, pandas have been remarkably unsuccessful. As of 1997, only six zoo males worldwide had copulated successfully to produce young. What's the problem? Ronald Swaisgood and his associates working at the Wolong Giant Panda Breeding Center suggest that giant pandas need to be exposed to appropriate odors at appropriate times in order to breed. Giving males access to the odors of estrous females may reduce aggression and heighten libido. Familiarizing a female with a particular male's scent can influence mate preferences and increase sexual behavior. They conclude that pandas probably need a gestalt of odors from many body sources to motivate and synchronize sexual behavior.

In the wild, giant panda mating is usually not a matter of a single female and male getting together in a bamboo thicket. The calls of a mating pair attract one or more other males. The Wolong Giant Panda Project found that as many as five males may congregate around a female and more than one may mate with her, the dominant male doing so first.

A critical consequence of the giant panda's low-energy diet and its minimizing the expenditure of energy is manifested in modification of various reproductive parameters, such as gestation length and litter size and weight. Gittleman concluded, "The greatest differences between the pandas (giant and red pandas) are in early neonatal characteristics. Neonates of the giant panda are extremely small relative to the mother. . . . Combined birth weight and growth rate revealed that both pandas are dramatically different from other carnivores." One of the consequences of eating a diet low in energy is small, slow-growing young—young that are vulnerable to a number of threats.

A CUB IN TIME

Seeing a newborn giant panda next to a robust, bamboo-crunching adult is startling. And this is a critical issue for giant pandas. A tiny baby panda evokes a feeling of compassion for a rare animal completely vulnerable in a rapidly changing world. That was one reason the World Wildlife Fund, for example, adopted the giant panda as its emblem.

Along with their slow growth rate and small litter size, giant pandas have low reproductive rates. They cannot compensate for any excess in mortality by producing more babies. Infant growth and development and mother-infant relationships in the giant panda have now been carefully documented in the Qinling Shan by Pan Wenshi's group, particularly by Zhu Xiaojian and Wang Dajun of Beijing University in collaboration with Donald Lindburg and Karin Forney of the Zoological Society of San Diego. For the first time, they followed the entire sequence of courtship, copulation, and

At between five and six months of age, cubs begin to climb and spend most of their time playing and resting in trees.

cub-rearing of individual females in the wild. They were able to do this by following radio-collared individuals, some habituated to their presence. Nothing comes easy where giant pandas are involved, and it took a decade of strenuous fieldwork to gain these critical insights into the early life of a giant panda.

In the study area, pregnant females descended in August from their usual summer home range at elevations of about 8,200 feet to between 5,900 and 6,600 feet. This was six weeks or so before other giant pandas descended to their winter home ranges. Large hollow conifers, like the ones females use as dens in Wolong, were absent from the Qinling Shan because of past logging. Instead, females used rock caves, located in the lower portions of their home ranges, as birthing sites. Females in Wolong did not descend to lower elevations to give birth but remained above 6,500 feet.

Females gave birth, usually to a single young (or, if twins are born, only a single young ever survived), about 139 days after mating. There was considerable variation in the giant panda's gestation time, although most of the births take place within a relatively short time span. Thus, the variation in the interval between mating and parturition that is made possible by delayed implantation may be a mechanism for avoiding late-season

A wild female panda gives birth to a single cub every two to three years, and may successfully raise only seven or eight cubs in her lifetime.

births and consequent exposure of cubs to severe weather.

The cave dens had little bedding but were a few degrees warmer than outside, important when the temperature is dropping as winter approaches. The pandas sometimes placed branches and sticks at the cave mouth, perhaps to hide the cub or prevent it from crawling outside the cave. Females used these cave dens for the first 100 days or so of their cubs' lives; then they moved them to lairs in dense patches of bamboo and fed nearby. In another two months, cubs climbed into trees where they spent much of the day, resting, exploring, and playing.

At birth, cubs are tiny, weighing only about 3½ ounces, or about 0.12 percent of the mother's body weight! (A proportionately small human baby would weigh 2 to 3 ounces.) They are pinkish, with short white hair sparsely covering the entire body. After eight days, the skin turns gray in areas where black hair will eventually grow. After three weeks, they have the typical black-and-white markings of adults. Their eyes are slow to open and are only narrow slits at forty to forty-nine days, half open at seventy-two days, and completely open in a few more days. This is when their first canine teeth appear; incisors follow at ninety days. The cubs' first attempts to crawl are at about sixty days, but they are unable to walk about efficiently until three to five months of age. In the first two weeks of life, the cub is highly vocal and squawks when slightly disturbed or seeking a nipple.

During the first two weeks or so of the cub's life, the mother remains with it constantly. From three to six weeks, she stays with the cub between 69 and 79 percent of the time, but after this she was away feeding most of the time.

Why this period of postpartum fasting by the mother when there is an abundant supply of bamboo nearby that she could eat? The most plausible answer is that the mother's presence is needed to protect the tiny naked baby from hypothermia, and from predation by a golden cat, dhole, or yellow-throated martin. The cub also suckles six to twelve times a day. The cub's dense pelage develops rapidly, roughly coinciding with the mother's first departures from the den. From birth to six weeks, the cub rests on the mother's body, cradled in her arms as she sits semi-erect or in a rocked-back position. If it moves, the mother carefully picks the cub up with her mouth and places it on her body. After about sixty days, the mother lays on her side with the cub resting quietly next to her.

Pandas become sexually mature at five-and-a-half to six-and-a-half years of age, and the subsequent interval between births is usually two years, although some panda mothers wait three. How many cubs make it to this age? This is a key statistic for the giant panda's future, but the information at this time is scanty. Survival data from zoo animals are not a reliable guide to survival rates in the wild.

There is a general impression that the low survival rates in zoos mimic the situation in the wild. Lu Zhi, China's foremost panda biologist, emphatically asserts this is not the case. Pan Wenshi, Lu Zhi, and their team working in the Qinling Shan found the infant mortality rate was about 40 percent. The giant panda's reproductive rate in these mountains was estimated to be about the same as in some populations of North American brown bears. A female may raise seven or eight surviving cubs in a reproductive life span of nineteen to twenty years. But we do know that any mortality factor that reduces cub survival is particularly threatening because giant pandas, already on the nutritional edge, cannot compensate by increasing litter size.

Twins occur in about half of births, but there are no records of two surviving in the wild. It seems impossible that a female could bestow the same intense level of care on two young that she does to rear one. Why she gives birth to twins about half the time is cause for continued speculation among giant panda specialists, but it is clear that tiny babies need much maternal care. Giant panda mothers invest all their energy in rearing a single cub through its first weeks. Between one and two months of age, a giant panda cub gains about 2 ounces a day. It grows faster than a single black bear cub, but then a female black bear is usually rearing two or three cubs, not one. But giant panda cubs grow at a rate equivalent to or more slowly than cubs of large-bodied cats and other bears. From a birth weight of about 3½ ounces, a cub weighs about 13 pounds when it leaves the den after about 100 days. At this age, its pelt has developed to the extent that it can tolerate the winter cold that has set into its mountain home.

The fat content of giant panda milk, while lower than that of superfat polar bear milk, is still high, about equivalent to American black bear milk. Producing this milk requires a lot of energy. Larger-bodied animals can store more fat than smaller-bodied ones. Wild female giant pandas have small fat stores in the fall, compared with other bears going into hibernation, but stored fat helps giant panda mothers get through the fasting period. After fasting during the young's first few weeks of life, the female has few options to improve her nutritional condition, given the low energy available in bamboo.

We've reviewed how a panda's life is about balancing the low quality of food with the quantity of food it can pass through the digestive track. Digestive efficiency strongly influences the cost and behavioral energetics of lactation. In mammals generally, caloric intake greatly increases during lactation: by 35 percent in black-tailed deer to as much as 149 percent in fox squirrels. A female giant panda can only increase bamboo intake a little. She could, but does not, shift to or add other foods. A slowly growing single cub is thus the option both red and giant pandas have chosen.

Cubs begin to eat bamboo leaves at five to six months and are fully weaned by eight months to one year, when they finally become grass-eating bears. They weigh about 80 pounds at a year old. They remain with their mother for two and sometimes three years before they strike out completely on their own.

After becoming independent, most but not all young appear to settle near their mothers. Some young females make long excursions away from their birthplace and settle elsewhere. That young females tend to disperse is unusual among mammals, and, as do all things giant panda, this intrigues their watchers. But this is also a critical feature to understand in order to develop conservation strategies for them. While dispersing, these young giant pandas venture out of their bamboo thickets. This makes maintaining or restoring corridors connecting their core reserve areas essential.

WHAT GIANT PANDAS NEED

"We know what the panda needs," Schaller wrote, ". . . a forest with bamboo, a den for its young, and freedom from persecution." We know that the giant panda is not a highly resilient species, but the activities of people in its range are the root cause of the giant panda's plight, not its unique adaptations to a diet of bamboo. Our conservation challenge is to ensure its ecological needs are met in its spectacular mountain landscapes. If and when we do, the giant panda will persist. The panda has not "had its day," as some say. The giant panda is a long-term survivor, and it is up to us to decide whether the future will contain a grass-eating bear.

As ambassadors for wild pandas, Mei Xiang and Tian Tian at the National Zoo may help ensure this species' survival.

5

THE PANDA'S UMBRELLA

We looked dubiously at the crumbling concrete stairs that led to the second story of the abandoned barn, then started to climb. If an 800-pound takin could negotiate the steep steps, then so could we. Shreds of matted dirty golden fur and piles of marble-sized pellets on the barn's second floor confirmed what we'd only half-believed: takin really were finding shelter in the attic of the derelict structure.

LT QXX

Previous pages: Animals that share giant panda habitat sometimes find shelter in a barn abandoned when farmers and loggers were moved out of Tangjiahe Nature Reserve.

Previous pages: Efforts to save giant pandas in the wild help other endangered species in their range, as well protect the watersheds that support human agriculture.

Large hoofed mammals called takin are among China's wildlife treasures, many of which live under the panda's umbrella in Sichuan Province.

Right: About 12,000 species of plants live in the region that is home to giant pandas.

A mile or so from the barn, which slouched not far from the headquarters of the Tangjiahe Reserve, the footpath along the river dwindled away. Only a takin trail snaked along the rugged valley slope. Scrambling clumsily, we saw clearly why no takin would retreat from a set of stairs. Despite their massive size, thickset build, and gawky appearance, takin tiptoe up and down the most precipitous of tracks and leap from crag to crag like ballet dancers across a stage. Takin can even balance on their hind hooves to nibble on vegetation 8 feet above ground. This animal's generic name—*Budorcas*—reflects this contradiction between ungainliness and grace: its translation is "ox-like gazelle."

The takin is just one of the spectacular species that share the giant panda's remote landscape, if not its celebrity. There are many more, plants as well as animals: the strangely beautiful golden snub-nosed monkey; Thorold's deer, which sports huge antlers; the glittering golden pheasant; the ethereal dove tree. China officially considers these "rare and precious." Leopards, clouded leopards, dholes, musk deer, and Asiatic black bear make appearances, too. While giant pandas are big stars, these less acclaimed animals and plants benefit from their light.

WILDLIFE HOTSPOT

The first Western natural historians began to explore the rugged high mountains that rim the Sichuan Basin—home of the giant panda—fewer than 150 years ago. They were

stunned with what they found: almost every plant and animal they collected was new to Western science. Today, scientists still recognize this region for its natural riches. In 1998, Conservation International identified twenty-five different regions around the world as being the highest priority for conservation action and investment. These hotspots possess high biological diversity and high levels of endemism—that is, species that are found there and nowhere else. The hotspots themselves are also considered fragile, and highly threatened. The endemic giant panda's home in central China is part of the "Hengduan Mountains of South Central China Hotspot."

The Wolong Biosphere Reserve is the best studied of this region's protected areas. Wolong means, auspiciously, "resting dragon." Among the real wildlife are some 93 species of mammals, 275 species of birds, 20 reptiles, 17 amphibians, 9 fishes, and 4,000 plants. For comparison, Great Smoky Mountains National Park in the eastern United States, which is the same size as Wolong, supports about 62 mammal species, 230 birds, 35 reptiles, 34 amphibians, 85 fishes, and about 2,000 plants. The Great Smoky Mountains are noted for an immense diversity of mushrooms, some 2,000 species in all. The southern part of the Hengduan Mountains boasts 4,000. Both of these protected areas fall into the World Wildlife Fund's Global 200 list of priority ecoregions; both are considered among the world's most diverse temperate forests.

The actual mix of species, both plants and animals, varies slightly from reserve to reserve, depending on topography and local climate. For example, there is potentially greater species diversity in the southernmost giant panda reserves such as Meigu Da Feng Ding owing to the infiltration of subtropical species. On the other hand, cold hardy Tibetan species predominate in the high elevations of the escarpment, called the Azure Wall, that rises west of the Sichuan Basin. Generally, however, all giant panda reserves estimate similarly large numbers of species.

The 4,000 different species of plants that grow in Wolong are a subset of about 12,000 in the region, including about 245 that are rare or endangered. The endangered dawn redwood, for instance, was discovered by Chinese scientists only in 1941. Found in the wild in just a few mountain valleys, it is considered a living fossil—the last of a once much larger group of trees. A unique deciduous conifer, it grows rapidly to reach 70 to 100 feet. Some seeds were planted in the United States and Europe in 1947, and today dawn redwoods are widely grown ornamentals. Several grow in Beaver Valley at the National Zoo. We drove along roads in Sichuan lined with dawn redwoods, their trunks whitewashed to blaze the track in the dark. On another road, in Meigu, we stopped to admire a rare and endangered dove tree. In the spring, this smallish tree's inconspicuous flowers are protected by bright white bracts (modified leaves) the size of handkerchiefs. Festooning the tree, they resemble a resting flock of white birds.

Père David revealed this gorgeous tree to the West, and it bears his name: *Davidia involucrata.* While this remarkable man will ever be linked in the public's mind with giant pandas, biologists recognize him for the countless other species he brought to scientific attention during his twelve years of exploring the edges of the Celestial Empire in the late 1800s. These include golden monkeys, serow, and other mammals, fifty-eight birds, about 100 insects, many snails and fish, and the Chinese giant salamander, one of many species named for him. The salamander may reach 100 pounds. Chinese once ate this beast as a delicacy. A similar giant salamander, the Appalachian hellbender, is a denizen of the eastern United States. At about 2½ feet long, the hellbender might fill a dinner pot, but Americans haven't acquired a taste for it.

Père David, a French Basque by birth and a Lazarist priest by vocation, was sent to China to save souls. Instead, with his real passion and talent for natural history, he undertook to save specimens of China's wildlife. He was lucky to have arrived in China just after Western gunboats compelled the isolationist Chinese rulers to open the interior to foreigners in 1860. Brave, perhaps foolhardy, and certainly arrogantly Western, he faced down bandits, illnesses, and other hardships to carry out his mission of collecting and naming animals. He was warned, for instance, against traveling to Muping, where he was to find the West's first giant

The missionary who revealed the existence of giant pandas to the Western world also gave this Chinese deer its name. This animal, now extinct in the wild, is called Père David's deer.

Lilies are among the many ornamental plants that hail from China; in fact many garden daylilies came to the U.S. from China within the last 100 years. Daylilies abound around Chinese homes, just as they do in parts of United States.

panda. He wrote in his diary, "I am told of thefts and murders committed by bands of brigands. But if I am to be held back by fears of this kind I can do no exploring, since the wild places, reputed to be the haunts of thieves and malefactors, are precisely the ones that offer the most in the way of natural history in China. . . . I shall be careful to keep my gun much in evidence."

He even risked the wrath of the emperor of China, and the death penalty, when he connived with guards to have two milou killed and their skins extracted (some might say stolen) from an imperial park south of Beijing. Now better known as Père David's deer, this animal had been disappearing in the wild for nearly 2,000 years. A herd lived in Nan Haitse, which was surrounded by a 43-mile wall. People were forbidden even to look into this royal enclave, but David first persuaded the guards to let him look, and then take, the strange deer.

Unfortunately, a flood in 1895 smashed walls in the park, and many deer perished. The last twenty or thirty were then butchered for food in the turmoil of the Boxer Rebellion in 1900. The last wild animal was reportedly shot in south China in 1930. But a handful of live deer, which reached England and France after Père David made the animal known, formed the nucleus of a successful breeding program that continues today. The National Zoo has been a major participant in this program for more than twenty years. In a strange quirk, when the London Zoo desperately wanted a pair of giant pandas from China in 1973, they exchanged them for four milou, animals the Beijing Zoo badly wanted. More recently, deer from Western zoos have been returned to China to launch a reintroduction program.

FAMILIAR EXOTICS

David was also a prodigious plant collector. Like the dove tree, many were named in his honor: butterfly bush, *Buddleia davidii;* a clematis, *Clematis heracleifolia,* var. davidiana; a lily, *Lilium davidi;* a peach, *Prunus davidiana;* and a photinia, *Photinia davidiana,* to name a few. Some Chinese still resent what they see as Western scientific imperialism, which not only took away plants, animals, and objects of art, but even the names of things. In Père David's Basque tradition, according to writer Mark Kurlansky, "naming something proves its existence." So perhaps giving plants and animals Western names could be seen as a sort of theft of their Chinese identity.

Many Western plant collectors followed in Père David's footsteps. British botanist Ernest Wilson, for instance, also known as "Chinese Wilson," visited China four times between 1899 and 1912. His adventures were no less harrowing than David's, but he managed to collect more than 3,000 plants, of which 1,000 were new to the gardens of the West. He was responsible for successfully collecting seeds of the famed dove tree for germination by an English nursery firm. His expeditions also bagged 3,135 birds and 375 mammals, al-

though he never happened upon a giant panda.

As a result, Western parks and gardens bloom in Chinese colors. Wilson called China the "mother of gardens." Many of our flowering ornamentals hail from these forests or are cultivars from China, including varieties of astilbes, lilies, roses, irises, camellias, and peonies. Wild peonies are threatened in China, where people use their roots as medicine. In fact, the peony's name in Chinese means "medicinal herb plant," and in ancient times giving someone a peony root was like giving a love potion.

Perhaps most stunning of central China's flora are the hundreds of species of rhododendrons that range in size from small bushes to tall trees. Meigu, the county of a reserve we visited, means "beautiful land," and in May this appellation is justified. Wild rhododendrons, flowering in colors from royal purple to lavender to pink to white, shine like jewels strewn among the verdant green of the forested slopes.

Other Sichuan plants evoke home to us because significant numbers of species pairs or larger species groups inhabit both eastern Asia and eastern North America. A French Jesuit first remarked on this pattern in 1716, when he discovered American ginseng near Montreal after reading about Chinese ginseng. Ever since, biologists have puzzled over the fact that such a large distance divides these closely related species, with no relatives connecting them. The phenomenon even has a name, the "Asa Gray disjunction," after the nineteenth-century

Graceful golden larches, cedars, spruces, and pines are trees characteristic of giant panda habitat.

Harvard University botanist who was a leading proponent of his friend Charles Darwin's theory of evolution by natural selection. Gray's description of the close affinities between the plants of eastern Asia and eastern North America was cited as key evidence of the theory. Scientists now suspect that forests stretched continuously across North America, Europe, and northern Asia, connected by the Bering Sea and North Atlantic land bridges. When rising sea levels submerged the land bridges and colder, drier conditions came to prevail across parts of this expanse, some plants disappeared from western North America and western Eurasia, but remained in East Asia and eastern North America. As a result, we felt as at home among the dogwoods, oaks, birches, firs, and ferns in the Sichuan forests as we do in the forests of Rock Creek Park in our own backyard.

Two seemingly typical North American plants followed a circuitous route to the United States, long after these climatic events, but before the plant explorers of the late nineteenth and early twentieth centuries sent their finds west. Peaches were already growing wild in what is now Georgia when the first English people were settling North America, but they hadn't been there long. Native to central China, peaches were taken to Persia, perhaps along one of the branches of the Silk Road, and from there Arabs took them to Spain when they ruled this country from about the year 700 to 1492 C.E. Spanish explorers later took them to Florida in the early 1500s, where they escaped cultivation and spread north to Georgia and beyond. The Cherokee rose, the most common wild rose in the southern United States, followed a similar path from China. (Plants did not move only in one direction; some American plants also made their way to China.)

West and central China harbors more sweet resources than just peaches: it is home to wild ancestors of many of our favorite fruits, including apricots, plums, cherries, and citrus, as well as vegetables and other crop plants. Tea, scallions, soybeans, peppercorns, and ginger also hail from Sichuan. There is a growing recognition of the importance of conserving wild varieties of domestic crops as a source of genetic material. Wild species may carry genes that confer resistance to disease, increase yields, or simply make food plants taste better.

Wild plants form the major ingredients in the pharmacopoeia of Chinese medicine. In all of China, there are more than 7,000 medicinal plants, only about 10 percent of which are cultivated. In the official *Pharmacopoeia of the People's Republic of China*, published in 1990, about half of the nearly 600 plant medicines are wild. And most rural people rely largely on wild medicinal plants. They also use wild foods to diversify their diets, from wild greens to hundreds of different kinds of edible mushrooms. In weeks of traveling and eating in Sichuan, we were served ten or twelve different kinds of mushrooms—all delicious—and any number of dishes of wild greens, including fern fronds, fiddleheads, and various fried leaves.

Left: Wild irises grown in alpine meadows and line paths through bamboo forests.

The elegant, endangered dove tree is found only in China.

Chinese people harvest thousands of kinds of medicinal and food plants, many of them from panda habitat.

The giant panda's bamboo forests are also noted for their abundance of stinkhorn mushrooms. These are the truffles of China, commanding similarly high prices in Asian markets.

Many other economically important plants also exist here. Walking through the forest in Tangjiahe Reserve, we happened on a midden of scattered clam shells that certainly could not have arrived there naturally. It turns out we were among lacquer trees, whose sap is used to make paint and oil. To collect the lacquer, people slash V-shaped gouges in the tree truck, then let the sap drip into the clam shell.

We found wildlife watching in giant panda country less rewarding than botanizing, something others have noted. The big-game hunting Roosevelt brothers wrote of the difficulty in seeing mammals during their 1928 expedition: "The parts of Yunnan and Sichuan through which we passed can by no stretch of the imagination be termed a big game paradise. It is no country for the inexperienced hunter nor for one who wishes to secure a large bag without undue expenditure of time and effort."

The stuffed specimens we saw in the Tangjiahe Reserve's small natural history museum tantalized us with a vision of what we might see alive in the wild. But on a cool November afternoon, takin were our only reward. Later, when we visited in May, the takin were too high in the mountains for casual hikers to meet. In about a dozen hours walking, we saw a porcupine, a tufted deer, a frog, a handful of birds, a couple of snakes, and some minnows. Only butterflies flitted in abundance. In Yele, Qianfoshan, and Meigu Da Feng Ding reserves, we saw no animals at all except butterflies and the odd insect, although admittedly we spent just a few hours hiking there. These are wild places that reward patient vigilance over days and weeks. But as in all such spots, around any curve in the trail, behind any bush, through any screen of leaves, you might confront some rare and wonderful creature.

Like so much of China's flora and fauna, snakes and other reptiles have not been very much studied.

COWS WITH GOLDEN FLEECE

Takin are such rare and wonderful creatures. A takin resembles a weird, impossible cross between a moose, a cow, and a bear. In fact, the takin's closest cousin is the musk ox. And, although sometimes called golden-fleeced cows or simply wild cows, takin are more closely related to sheep and goats than to bovines. Although Père David was the first Westerner to collect a takin, these fairly conspicuous animals aren't quite so mysterious as giant pandas. Marco Polo heard of takin as dangerous animals, and plant collector Ernest Wilson reported that local people, who hunted the animal with primitive weapons, believed it to be "both fierce and revengeful."

Takin inhabit rugged, mountainous terrain, ranging from 4,000 to 14,000 feet, where they break trails—more precisely, narrow ruts—in dense thickets of rhododendron and bamboo to travel between foraging areas. They migrate vertically, moving from high-elevation alpine habitats in the summer to forested valleys in the winter. Menus change with the seasons, too, according to what is available to eat. George Schaller, who observed these animals during his study of giant pandas in China, documented the catholic food habits of takin. "Takin eat almost any plant within reach of their broad mouths," he said, including the tough leaves of evergreen rhododendron and oak, willow and pine bark, horsetails, bamboo leaves, and a variety of more succulent, new-growth leaves and herbs. Takin in one herd ate twenty-seven different items while foraging through a single field.

Schaller also noted an unusual feature of the takin's social system. Groups of young takin—Schaller saw up to sixteen—are often tended by just one baby-sitting female while the other mothers forage and socialize. In general, takin society appears fluid. While females and youngsters live in groups, males are generally solitary but may associate with a group of females and young, or with other males. Groups of ten to thirty-five animals were seen, but sometimes these groups coalesced into a larger herd of 100 or more. A salt lick is often the magnet that attracts such large groups.

Even a large group of takin dotting a distant slope can easily be overlooked in the dense vegetation. Through binoculars at dusk, we squinted to resolve pale golden specks

Once hunted for their beautiful fur, golden snub-nosed monkeys are endangered. Now, Chinese law protects these primates as strictly as it does pandas.

into large furry ungulates on the cliff across the river. Only their measured movements gave them away. In this part of China today, predators able to take down an adult takin—such as leopards, dholes, and tigers—are scarce or have been extirpated. At higher elevations, snow leopards, with their adaptations for hunting prey in rugged mountainous habitat, may kill takin. The takin here are wary nonetheless, perhaps fearing the human predator.

Local people hunt takin for their highly prized meat, and poaching is reducing their numbers throughout the species' range. In Wanglang Reserve, which we visited in November 1999, we almost tripped over a poacher carrying the carcass of a takin through the forest not more than a few miles from headquarters.

Deforestation has also taken its toll on takin, as it has on much wildlife. Two of the four subspecies of takin are found only in central China, a range roughly overlapping that of giant pandas. One of the subspecies—the golden takin—is listed as endangered, while the Sichuan takin is listed as vulnerable by the World Conservation Union. Two other subspecies' ranges are in the Himalaya, including parts of Myanmar, Tibet, Bhutan, and India. Of these, the Bhutan takin is considered vulnerable, and the Mishmi takin is endangered. China extends the takin its highest level of legal protection: like the giant panda, the takin is officially a national treasure.

Although little known, serow are widely distributed in tropical and subtropical Asia, including the mountains of Sichuan. Serow are relatively small ungulates, weighing up to 300 pounds, with long tasseled ears, short stiletto-sharp horns, and short solid hooves. The Chinese name for the serow means "cliff donkey," but serow are related not to donkeys, but rather to North America's mountain goats and Europe's chamois. These fairly solitary animals live on cliffs and on brush-covered mountains between 3,000 and 12,000 feet, where they have been described as "adept at dashing wildly down impassable precipices." Serow are thought to be selective browsers on leaves and other plant parts, but few particulars are known about their food habits, or, indeed, about other aspects of their behavior and ecology.

Pandas are sometimes caught in traps set to capture muntjacs, small deer that make a good meal.

MORE GOLD

Another of China's endangered treasures is the golden snub-nosed monkey, which ranges from the high mountain forests of Sichuan to the Tibetan border. Alphonse Milne-Edwards, the Paris Museum of Natural History scientist to whom Père David sent his specimens for study, playfully named this lovely creature *roxellanae,* after the snub-nosed mistress of a Turkish sultan. One of just a handful of primates adapted to life in a temperate climate, these beautiful long-haired monkeys sport orange fur on much of their body and pale-blue patches above the eyes and nose. One observer wrote, ". . . their nose looks as if a bright blue butterfly was sitting with its wings open in the middle of their face. . . ."

The arboreal golden monkeys are leaf-eaters, but supplement their diet with fruit, seeds, insects, birds, and bird eggs. Large troops of up to 600 individuals have been reported, but where their habitat is disturbed, troops may number only thirty to 100 monkeys. The troops are subdivided into small groups composed of one adult male, about five adult females, and their offspring. Little more is known about the lives of golden monkeys. One well-known fact, however, is that they are vulnerable to extinction. Long hunted for their fur, which is fashioned into coats, and for other body parts used in traditional medicine, golden monkeys are offered China's highest level of protection, and hunting is forbidden. These primates, which are estimated to number between 6,000 and 14,000 individuals, remain threatened by continued habitat fragmentation and destruction.

DIVERSE DEER

Deer of several kinds find homes in central China. Tufted deer little resemble the familiar American white-tailed deer. Tufted deer stand just 25 inches at the shoulder, and the male's very small, unbranched antlers are almost completely hidden by tufts of hair on the forehead. Unlike most deer, the males also possess long, sharp canine teeth that extend like small tusks or fangs over the bottom lip. Male tufted deer probably use these canines in fights with other males, just as male muntjac—other small deer found in panda habitat—use their similarly long canines in fights over territory and females. Tufted deer live at altitudes up to about 15,000 feet in high valley and mountain forests.

Musk deer live in giant panda habitat, too. Five species of musk deer (some experts say four, others six) live in various parts of China; two overlap with giant pandas. Slightly smaller than tufted deer and muntjacs, musk deer lack antlers entirely, but their fangs grow up to 3 inches long—maybe a better name for them would be Dracula deer, despite their vegetarian diet. The word "musk" derives from an old Sanskrit word for "testicles," and refers to the glands enclosed in sacs found near the male musk deer's genitals. The scent of the secretions of these glands—musk—may attract females for mating.

People have voraciously hunted these deer for their prized musk for centuries to use in perfumery and in traditional medicine. So big was the market for musk that Isabella Bird, who visited Chengdu in 1896, wrote that the city "reeks with its intensely pungent odor." Although most musk used in perfumery is now synthetic, it is still an ingredient in 300 to 400 manufactured medicines used to treat conditions ranging from asthma to apoplexy. TRAFFIC, the international wildlife trade watchdog, estimates that China alone may use as much as a ton of musk annually. And because each gland contains only tiny quantities of musk, it may take as many as 100,000 male musk deer to meet this yearly quota. As a result of hunting and poaching, coupled with habitat loss, musk deer are considered near threatened, and efforts are underway to reduce both legal and illegal harvesting and trade of musk. Moreover, giant pandas are sometimes snared in traps set for musk deer, creating another hazard for the beleaguered bears.

Found only in a small area in central China and Tibet, the 300-pound Thorold's deer, also known as white-lipped deer, more closely resemble American white-tails but are adapted for life in high-elevations. Like takin, Thorold's deer can easily climb rocks and negotiate steep slopes. Males sport huge racks during the breeding season; their multitined antlers may extend more than 4 feet high and weigh 15 pounds. This species is considered vulnerable by the World Conservation Union, but it may number as many as 100,000 individuals. These deer are farmed in China, where their velvet antlers—the stage before the antlers turn bone-hard—are used in traditional medicine.

FANTASTIC FLYERS

Wakening early in the Wanglang and Tangjiahe reserves, we heard the calls of pheasants that contribute to the characteristic soundscape. The mountains of central China are home to nine of the world's forty-eight species of pheasant, including four species found nowhere else. These large, forest-dwelling game birds are best known for the often spectacular plumage of males. While female pheasants' drab markings in shades of brown and gray blend into the colors of the forest, male markings are like bright beacons, with bold patterns in

Glittering golden pheasants are among the nine pheasant species that live in central China.

Like this Impeyan monal, all but one species of pheasant are threatened. They are hunted for food and affected by habitat loss.

brilliant colors. The males of two of central China's pheasant species are among the gaudiest: Lady Amherst's pheasants, also known as flower pheasants, shine with patches of crimson, yellow, orange, blue, and metallic green; golden pheasants glitter with a long, bright-yellow crest and crimson underparts, and spots of green, blue, red, and yellow on their back and tail.

Other Sichuan pheasants include Temminck's tragopan and white-eared pheasants, as well as Reeve's long-tailed pheasant and the endangered Chinese monal. All the pheasant species, which with one exception live naturally only in Asia, are threatened. Good to eat, they have been intensely hunted throughout their range. As sedentary ground nesters and feeders, they are vulnerable to habitat destruction and the impact of free-ranging domestic livestock, such as goats, which compete with them for food.

Found only in the Daliang Shan region, where the southernmost population of giant pandas also lives in Meigu Da Feng Ding and its sister reserve in Mabian, Sichuan hill partridges are critically endangered. These quail-sized birds inhabit broadleaf forest and have suffered from deforestation, but it is encouraging that they appear to be living in once-deforested areas that have been replanted with broadleaf trees.

Across the border in the Qinling Mountains of Shaanxi Province, the last crested ibises in the world cling precariously to life in patches of pine forest. These beautiful wading birds were widespread until the mid-1800s, breeding in southeast Siberia, north-

The masked palm civet is one of the largely tropical species that occurs in the giant panda's temperate habitat.

east and central China, northern Korea, and Japan. But shooting and logging of pine woodlands reduced their numbers. Fertilizers and pesticides took a devastating toll, fouling rice paddies and wetlands where the birds feed on insects, amphibians, crustaceans, and fish. Crested ibises have disappeared in Siberia, North Korea, and most of China. Japan's last five birds were captured from the wild in 1980, but the breeding program failed. In 1999 and again in 2000, two ibises sent to Japan from China to revive the program produced a chick, bringing Japan's ibis population to four. Only in Shaanxi did wild birds hang on, and, thanks to Chinese conservation efforts, this population has grown from as few as seven to a few dozen today. Some are also in conservation breeding programs in China. In total, about fifty ibises remain.

PREDATORS

With such a banquet of potential prey, it's no surprise that many species of carnivore lurk in the forests of cen-

Red pandas share a name and, sometimes, space with giant pandas. They too are endangered in China and in other parts of their range.

tral China. Tigers have been extirpated, however, and leopards may be well on their way out. In four years of collecting leopard scats in Wolong, a team of scientists watched their numbers dwindle. They collected thirty-six scats in 1985, thirty-two in 1986, nine in early 1987, and none in late 1987 when their study ended.

But five other cat species still hunt in parts of the region. Small Pallas's cats with snub noses and lush golden fur slink among rocks at high elevations, possibly preying on small pikas, also called rock rabbits. Husky Eurasian lynx probably live on roe deer, while agile snow leopards stalk blue sheep at very high elevations. Clouded leopards have lived here in the recent past, and may still. Eight kinds of mustelids, including river otters, hog badgers, and tiny least weasels find niches here, too, as do palm civets, usually tropical fruit-eaters of which we saw abundant sign in Tangjiahe Reserve. Among the canids are two species of foxes and, maybe, wolves. And then there are dholes, reddish-furred wild dogs that hunt in packs like wolves do. They have been observed to attack giant pandas.

Apart from giant pandas, two more bears are denizens of this rich ecoregion: brown bears, called horse bears in China, and Asiatic black bears, called dog bears. Brown bears here, like brown bears in North America, are seldom seen, while Asiatic black bears, like their American black bear counterparts, can live among people, and among giant pandas. In Tangjiahe, black bears play the Yogi Bear role well, regularly raiding domestic beehives for honey. Some of their favorite haunts here are abandoned farmsteads where apple and peach trees still fruit. In Meigu and elsewhere, the black bears, along with red pandas and wild pigs, dig up tasty potatoes and munch corn on the cob. Farmers find these antics less than amusing and may bury explosives in meat to bait, and blow up, the bears stealing their sustenance. Giant pandas, also attracted to meat, are sometimes unintended victims of the booby traps.

Schaller and his colleagues examined how Asiatic black bears and giant pandas—both solitary, herbivorous carnivores of similar size—can coexist in the mountains of western Sichuan, specifically in the Tangjiahe reserve. Their most important finding was that the two species do not compete at all for food. Both eat bamboo shoots, but only when these are abundant in spring and summer. Otherwise, black bears eat a diverse, high-quality diet of small succulent green plants in the spring; acorns in the fall; berries and other fruits in spring, summer, and fall; and then nothing during their winter hibernation. In contrast, giant pandas stick monotonously to low-quality bamboo—shoots, leaves, and stems depending on the season—year-round. Black bears are like people who invest in high-risk stocks, living high when the market is good, crashing when it's bad. Giant pandas resemble the cautious folks who put their money in savings accounts, always gathering the same small but steady earnings.

THE OTHER PANDA

Red pandas will likely always be overshadowed by the black-and-white giants that eat the same food, live in some of the same places, and share a name. In fact, this species was originally *the* panda, named for now-obscure reasons in 1825 by the French scientist Frédéric Cuvier. Only after zoologists erroneously concluded that the big black-and-white Chinese bamboo-eater was more closely related to it than to bears did it suffer the indignity of becoming known as the lesser panda. In our more sensitive times, however, the preferred moniker is red panda, although another name, fire cat, is more evocative. Arguably, this panda, with its cinnamon red fur, long bushy tail, and sweet face with big eyes and wide ears, is the more attractive of the two. But the giant has stolen center stage.

Red pandas live in the mountains of Sichuan and other parts of south central and western China as well as in the bamboo forests of the Himalaya in Myanmar, Assam, Bhutan, and much of Nepal. Along with the bamboo that forms a significant portion of their diet, red pandas eat fruit, seeds, acorns, and probably insects and small animals if they can catch them; in zoos, they are happy to eat some meat. Solitary and retiring, they spend much of the day sleeping and sunbathing in trees, descending to the ground at night to forage. Red pandas are not immune to the dangers that threaten giant pandas and other highly specialized species in their rapidly changing Asian landscape, such as deforestation and hunting. The species is considered endangered in China and is believed to exist only in small numbers elsewhere. Fortunately, in the last thirty years, more than 300 have been born in zoos worldwide, including many at the National Zoo.

BAMBOO RATS

Finally, a few words about the bamboo rat, by name alone a loser in any bamboo-eater popularity contest. These burrowing rodents weigh about 2 pounds and have stout bodies with short legs ending in powerful digging claws. They sometimes leave their burrow, cut off a bamboo stem, and drag it back underground. Alternatively, from the safety of their subterranean lairs, they may sever a stem where it joins the rhizome underground. Schaller recounted his amazement when he saw a bamboo stem appear to grow in reverse when a bamboo rat pulled it from below ground. While not endangered, these animals are sometimes dug up and eaten by people.

PEOPLE IN PANDA LAND

Driving between five different giant panda reserves, on two trips that took us around the mountainous rim of the Sichuan Basin, we traveled through isolated farms and small villages, through growing towns and bustling but smallish cities like Pingwu and Meigu. Outside of Chengdu, and off the few hundred miles of paved highway, this region feels as rustic and remote as Idaho, or the past. But many of the farmhouses perched on defiles sport satellite dishes capturing the television signals of CNN and MTV amid the drying garlands of red peppers and pale yellow corn on the cob. Nowhere is truly far away anymore, so we can no longer think of wildlife and wild lands surviving in some pristine, people-free place.

From bamboo rats to dove trees, the fate of all of the wildlife and wild lands of central China is tied to the future of giant pandas, which act as umbrellas to shelter the lesser-known inhabitants of the region. The giant panda's umbrella also shields watersheds and the forests that help control erosion and flooding. So saving wild pandas is not only about saving cuddly creatures. It is about improving the lives of the millions of people who depend on natural resources, too.

The fight to save giant pandas will be won—or lost—in the middle landscapes where people and pandas live together.

6

CHINESE SHADOWS

The pastoral landscape of the Sichuan Basin is a brilliant tapestry of golds, greens, and browns, splotched with the vivid red and blue blouses of the women working the land. In villages and small settlements, homes are roofed with rust-colored clay tiles. Doors are framed in colorfully painted borders, and even the seemingly humblest of homes display flowering plants in pots. In kitchen gardens and larger plots, corn, potatoes,

beans, and other greens are growing. Patches of flowering rapeseed, the source of canola oil, gleam yellow. Immense paddies of jade-green rice predominate, while smaller, terraced plots are irregularly shaped to fold into the low hillsides and follow the contour of the land. This is the vision of Sichuan that earned the province its reputation as Heaven's Storehouse, and is what we saw driving from Chengdu to Emei in late May.

Our Chinese hosts had warned us to expect a hard day of travel. But this stretch had been easy enough, on a major new toll highway connecting the capital and the self-proclaimed "Top Tourist Town of China." Two thousand years ago, the first Buddhist temple in Sichuan was built on the top of 10,000-foot Mount Emei, and additional temples followed nearby; the area became one of Buddhism's holiest sites. It is also an area of great natural beauty, and parts of it are in protected areas supporting takin and other wildlife. Pandas were seen here until 1948.

But from Emei, the road starts to wind uphill and rapidly deteriorates to one unpaved lane hanging onto the edge of the cliffs. Waterfalls like tears stream down the sheer rock faces. Rice paddies disappear, replaced mostly by corn and potatoes, the staples of higher-elevation, steep-slope agriculture. It takes more than an hour to travel the 36 miles from Emei to Ebian, and from Ebian it takes another four-and-a-half hours to go the remaining 100 miles to our destination of Meigu. Two days later, continuing south, it took another four hours and 95 miles of rough paved road to return to the lowlands. We understood the old saying that it is harder to get to Sichuan than it is to get to Heaven.

This area is home to one of China's official ethnic minorities, known collectively as Yi, who are actually made up of several culturally distinct groups. People are clearly poorer here than in the lowlands, and their lives harder. The landscape looks tired, with little of the lush pastoral glow of the Sichuan Basin or the Min Shan where we'd traveled a few days before. Only the people are bright here—women in traditional Yi dress are as colorful as a flock of pheasants.

Rocks and boulders litter the ground; clearing them for cultivation must be a terrible and continual burden. Narrow goat trails climb straight up the mountainsides, and so do the people who farm terraced plots right up to the top of the ridges. Homes lack electricity, so fuel wood is stacked high, as is cut bamboo, which is used to fashion tools and fences. The lovely red-tiled roofs of richer areas are replaced with wood or tar paper anchored with rocks. What land is not farmed is overgrazed by free-ranging goats, pigs, and water buffalo. There is little of the reforestation on steep slopes that we'd seen elsewhere. Children abound; while most Han Chinese couples follow the one-child rule, minority families are permitted three children, and the difference is strikingly evident to passersby. Yet at the top of a ridge, at about 7,400 feet, we suddenly entered a rich stretch of mixed broadleaf and

Previous pages: Highly productive agriculture earned the Sichuan Basin its reputation as a Heaven on Earth.

Left: Little deciduous forest remains in Sichuan.

Yi women in traditional finery welcome rare visitors to their inaccessible autonomous region, which is also home to the southernmost population of giant pandas.

Population growth rates remain high among China's ethnic minorities.

Right: The worn clothing of these Yi people reveals the poverty of ethnic minorities living in and around giant panda habitat in the mountains of Sichuan.

coniferous forest with dense patches of bamboo. Giant pandas live here.

In this brief journey through space in modern Sichuan are echoes of a journey through time, with clues to changes in land and landscapes in China's environmental history. People have lived in south China for about as long as there have been people. It is very likely that human activities once may have created giant panda habitat. Disturbance from the kind of shifting cultivation and forest gardening that have characterized millennia of human history promotes and sustains bamboo, for instance. Modern human behavior, however, has largely been in conflict with the interests of pandas and other wildlife. Moving from competition for resources, including food, space, and forest, between people and giant pandas to cooperation to ensure enough resources for both will not be easy. But a first step toward that goal is understanding how people and pandas got to this crisis point.

At the end of the Pleistocene, about ten thousand years ago, giant pandas were distributed over a much wider area of south China, but the historical record of giant pandas is sparse. Wen Huanren and He Yeheng summarized the record, noting a rapid decline of pandas after the end of the Pleistocene. They also describe the nineteenth century as a critical period in the decline of the remaining pandas, largely due to rapid population growth and subsequent land clearing for farms. More generally, a brief history of environmental change across south China—wherein giant pandas gradually retreated to their present stronghold—offers glimpses of the past as prologue.

OF RICE AND PANDAS

China is roughly divided by east-west running mountain ranges, including the Qinling range where the northernmost giant pandas survive. North of the divide is the Yellow River watershed, a cold dry area where wheat dominates agriculture. South of the divide is the Yangzi River watershed; here, wet rice dominates agriculture. In historic times, the range of the mountain giant panda fell in the wet-rice region south of the mountain divide, although they were probably confined by their diet and cold-adapted physiology to moist areas at higher elevations. The history of south

The continual expansion of wet-rice agriculture in China in the last 2,000 years has displaced both elephants and the lowland giant panda.

Right: Corn and potatoes are among the many crops imported to China from the New World. These crops enabled people to farm at high elevations, where rice and wheat do not grow.

China is primarily a tale of the replacement of various indigenous peoples engaged in hunting and gathering, shifting cultivation, and, in some case, pastoralism, by Han Chinese engaged in sedentary, intensive wet-rice farming. Han now comprise 90 percent or more of the Chinese population. In most areas, the conquest was accomplished by the assimilation, not annihilation, of other peoples.

In *The Retreat of the Elephants,* environmental historian Mark Elvin describes the impact of this process on elephants, which were once common over much the same broad geographical areas as giant pandas, and vanished in a roughly similar pattern. Some of this decline can be attributed to climate change; for instance, southeastern China appears to have become drier after about 1230 C.E. But Elvin writes, "The pattern of withdrawal in time and space was, so to speak, the reverse image of the expansion and intensification of Chinese settlement. Chinese farmers and elephants do not mix." In some cases, what Elvin calls the war between human and elephants was direct—hand-to-hand combat if you will—in which elephants were killed as crop thieves, captured for work and ceremony, and hunted for ivory and for their trunks, which were gourmet food items. But what pushed elephants to the very edge in China—a few are holed up in southwest Yunnan on the border with Myanmar—was the steady replacement of low-elevation forest with farms. Elephants cannot live without forest or on high-elevation steep slopes.

Although this process had begun as early as the first century or before, historian Robert M. Hartwell talks of a diaspora from north to south China between about 880 and 1150 C.E., thanks to political turmoil in the north and improved varieties of rice. He notes, "Before 750, two-thirds of the Chinese lived in the dryland wheat farming areas of the North. By 1150, two-thirds of the inhabitants . . . farmed fields in the irrigated paddy agricultural areas of the South. This transformation was accompanied by at least 100 percent overall growth in population from about 70 million to 150 million by 1150." The population crashed between 1250 and 1400, due to famine, wars, and epidemics.

Numbers recovered by 1580, for a "regional distribution and total population that was almost an exact replica of what it had been in 1150." There was another population crash in the mid-seventeenth century but, from then on, population size generally continued to grow. Today, China's population is at 1.2 billion.

Richard von Glahn's *The Country of Streams and Grottoes* documented, as the subtitle states, the *"expansion, settlement, and the civilizing of the Sichuan frontier in Song* [960–1275] *times."* His description echoes Western European expansion into the New World:

> For the Han immigrant, the frontier appears as a barren, desolate wilderness which impeded . . . normal habits of life and work. Yet the frontier was not a pristine landscape. The native people subtly altered the ecology of their habitat to meet their own basic livelihood needs. Although divided into myriad distinct societies, the native peoples created remarkably similar forms of livelihood under similar ecological conditions. Indeed the southwestern frontier was the meeting point of two distinct civilizations, the indigenous civilization of the forest . . . and the civilization of the plain, borne by Han immigrants intent on refashioning the frontier into a typically Han world.

Immigrants favored river valleys and plains, which are ideal for growing rice, and these were the first to be deforested. Some of these areas, includ-

Scenes such as this appear eternal, but in fact change has been a constant in China's rural landscapes.

Corn, or maize, arrived in China about 450 years ago. It is now a staple among people living in the mountains of central China.

ing the Sichuan Basin, had been cleared for agriculture since antiquity. This probably spelt the demise of the warm-adapted lowland panda. Other areas remained wilderness much longer. Unlike elephants and the warm-adapted pandas, cold-adapted giant pandas in their mountain lairs were probably not greatly affected by lowland deforestation until it was so complete that all dispersal routes between populations were cut off. Farmers didn't seem to have been bothered by giant pandas raiding crops—pandas were like shadows, seldom seen, rarely remarked on, even in the voluminous and detailed records the Chinese have been keeping for millennia. Neither did people ever seem to hunt giant pandas aggressively and systematically, as they did elephants and tigers.

But when people began climbing into the hills and mountains and cutting down the forest, pandas began to encounter grave trouble. Two broad trends sent people up the mountain slopes: an ever-growing demand for timber for construction and fuel, and a new ability to farm ever higher in the mountains, thanks to the introduction of New World crops.

COLUMBIAN CONNECTION

One afternoon in Chengdu, we visited the parkland site of Du Fu's Thatched Cottage. Du Fu is widely regarded as China's best poet, and the thatched cottage is a tribute to him. The cottage is not original—after all, Du Fu lived here in the 760s C.E.—and it is probably more spacious than the

恭喜發財
財源廣進家家樂

home of a poet in exile was likely to have been. But it looks rustic enough and is set amid a bamboo grove like ones he mentioned in his verses. To lend the cottage further verisimilitude, bundles of corn and chili peppers are hanging to dry from a wooden beam over the porch. We saw the same red and yellow garlands everywhere in the countryside. But Du Fu wouldn't have seen them in 760. In fact, nearly 800 years would elapse before these and other New World crops appeared in the Asian Old World.

Portuguese traders seeking silk and porcelain were probably responsible for introducing corn, chili peppers, and other New World crops, including white potatoes, sweet potatoes, peanuts, and tobacco, to south Asia. Exactly how they reached various parts of China is not well known for most areas. Some implicate Jesuit missionaries, but simple trade and cultural diffusion along established Asian trade routes may have worked as well. Environmental historian Robert Marks found a gazetteer record of sweet potatoes arriving in a southeast China county in 1580, "when one Chen Wen brought back some . . . from Annam (present-day Vietnam)." When these crops got to various locations is not usually so precisely known. Their impact on China's environmental history, however, was as profound as the chili-fired heat of Sichuan cuisine.

Corn arrived in China sometime before 1555, perhaps twenty or thirty years before, according to the classic 1955 paper on the subject by Ho Ping-Ti. By the end of the eighteenth century, it was the primary food crop of southwestern China. Corn grows in poor soil at high elevations where wheat and rice cannot and, unlike these crops, can be usefully planted in rocky, pocket-sized terraces. In parts of Sichuan, you see patches of earth the size of a bath towel hanging between the edge of a road and a sheer slope that support a dozen plants. And even at low elevations where rice rules, smaller patches of corn sprout between paddies and in kitchen gardens.

The rapid acceptance of New World crops may have helped fuel population growth across China by providing insurance against poor rice or wheat harvests. Along with increased double cropping, more intensive use of fertilizer, and other agricultural changes, this helped the population of China to about triple between the late seventeenth and early twentieth centuries. Other historians argue that it was population pressure and a growing shortage of arable land that spurred acceptance of crops that made mountain slopes profitable to farm.

The introduction of corn let people intensively farm the mountains, especially in southwestern China. To do that, however, people had to cut down the trees. Furthermore, growing and harvesting corn requires little labor compared to rice and wheat, freeing people to clear yet more forest to grow cash crops, such as peanuts and tobacco, for the market. Sweet potatoes offer similar advantages. Potatoes came to China later, in the mid-seventeenth century, but they, too, grow in marginal land and at high elevations,

Left: Central China's forests have been cleared and allowed to regenerate repeatedly since people first began to live here.

The natural beauty of bamboo has inspired artists and poets for millennia.

Clearing of coniferous forests in panda habitat leads to hardwood forests dominated by birch. Pandas prefer the bamboos that grow in the shade of conifers to those growing under birch.

Posing few problems for people, pandas have never been persecuted like elephants and tigers have been.

and were quickly adopted. Robert Marks, who in *Tigers, Rice, Silk, and Silt* emphasizes the importance of expanding market economies in environmental change in China, notes that once they could substitute corn or sweet potatoes grown on marginal land for rice, people could convert high-quality paddy land to cash crops like silk, sugar, and fruit.

Kenneth Pomeranz has recently questioned the traditional view that "the goal of those moving up the hillsides was to find some place to produce more calories: deforestation occurred in order to clear an obstacle to creating new fields." He argues that mountain forests weren't cut down to grow corn, but that corn was grown to cut down mountain forests, at least in the southeastern provinces he studied. Corn, he wrote, "provided a much cheaper and more reliable way of feeding loggers during the two to three years that it typically took them to finish clearing a hillside than could shipments of grain from the lowlands." With timber increasingly scarce in the nineteenth century, the availability of high-altitude crops made it financially worthwhile to invest in logging operations in remote and difficult terrain. On a similar theme, historian Leong Sow-Theng suggested that migrants might move to highlands when markets for timber or cash crops were good, then revert to growing food crops when demand fell.

Pomeranz also suggests that much of the damage caused by farmers moving from lowlands to highlands was due to people using the same farming techniques on slopes or at high elevations that they used on lower, flatter land, rather than employing the more appropriate techniques of the long-term local inhabitants. This has also happened in other parts of the world, where newcomers created havoc by insisting on using the farm methods of home rather than those of the indigenous people. This idea remains to be explored in panda landscapes in China.

In Sichuan, intensive wet-rice agriculture has long occupied the plains and pushed into the lowlands, while indigenous pastoralists grazed their livestock in the alpine meadows above tree line, sometimes burning forest to expand grazing downward. A few shifting agriculturists, and giant pandas, moved through the middle. With the advent of corn and thus the opportunity for settled farming, a big squeeze began. The middle has slowly disappeared.

There were other pressures on forests in Sichuan as well, dating from antiquity. Apart from rice and other products, notably silk and tea, Sichuan was known for its salt. Once an inland sea, the basin is rich in deep deposits of salt, and salt mining was once a huge industry. In a complicated process, bamboo tubes were used to bring brine up from deep wells. Wood from the then-abundant forests fueled the fires that boiled the brine to extract salt. Another interesting source of demand for wood and bamboo for making paper was bookmaking. Sichuan at the end of the first millennium was the first area to produce significant numbers of printed books and, later, the world's first printed money.

In the late nineteenth and early twentieth centuries, Sichuan was a major center of opium poppy cultivation for trade to the West. Land was cleared to plant poppies, which reportedly exhaust the soil, so new plots had to be cleared. A Chinese colleague told us that panda habitat in several areas was once planted in poppies; George Schaller was told that about a larch-covered hillside in Wolong Reserve.

Mapping historical records of giant pandas, a cluster of data points appears in eastern Sichuan and to the south in Guizhou. They are recent records, from 1603 to 1865. According to Elvin, the Han arrived here relatively late, and some of the local inhabitants, the Miao, who themselves had displaced earlier peoples during the first millennium C.E., resisted Chinese government control. By all accounts, this was a wild frontier area, thick with bamboo in immense forests. Even though people had long lived there, Elvin quotes a seventeenth-century Chinese observer who saw this landscape as a primal wilderness untouched by human hands. Monkeys, tigers, deer, and other animals, including, apparently, giant pandas, thrived there. The allure for the Chinese was largely rich deposits of cinnabar, the only common source of mercury, which was used in various industrial processes.

Both the Chinese government and the Miao practiced arboricide in their battles with one another. Having moved up the mountains, the Miao cut huge trees to block old mountain roads, while government forces cut huge trees to build new mountain roads. Worse, hillside forests were deliberately burned to remove cover and expose rebel strongholds. In 1850, a government official wrote, "We have driven away the wolves and foxes, and opened up the rocky mountains and the weedy wastes for farming." Nonetheless, warfare continued until the early 1870s. The last mention of giant pandas in this general area was in 1865. Here, it seems, warfare interacted with expanding agriculture, itself a mechanism for increasing Chinese settlements, to eliminate the giant panda's mountain habitat.

Pomeranz told us that this warfare by deforestation occurred over and over, especially during the White Lotus

If logs weren't being harvested in Yele Giant Panda Reserve, they would be lost when the area is flooded following dam construction. But how this affects pandas is unclear.

Rebellion of 1796 to 1804 and the Taiping Rebellion of 1851 to 1864. The White Lotus Rebellion was centered in the mountains straddling the borders between Sichuan, Hubei, and Shaanxi provinces, in core giant panda habitat. The Taiping Rebellion began in Guizhou and ended with the defeat of the rebels in Sichuan and elsewhere, so it too swept through panda country. Less than a century later, in fighting before, during, and after World War II, various warring factions cut down and blew up forests in Sichuan, as well as in other parts of China.

Wen Huanran and He Yeheng also reported that after forest was cleared for farming, farmers burned and cut down still more forest to destroy the habitat of crop predators like wild pigs, badgers, and monkeys.

The reasons for the kind of environmental change that led to the giant panda's decline are variations on a common theme. Whatever the particular situation—expanding agriculture, increased grazing, ecological warfare, salt mining, or a growing population's need for wood—deforestation followed.

FLOODS AND FOREST

Deforestation has proceeded inexorably in China since the dawn of history. Forests have been cut, and have grown back or been replanted over and over. In *Patterns of China's Lost Harmony,* geographer Richard Louis Edmonds quotes a source in Chinese: "[E]ven in the rugged parts of China, in areas that look like untouched wildernesses, subtle evidences of the human presence occur." One of these signs is bamboo. Edmonds recounts many documented episodes of deforestation in Chinese history. For instance, the Yellow River and Wei River valleys were severely deforested between 221 B.C.E. and 220 C.E. to build dynastic capitals (twice; the first capital was burned) and meet other needs near what is now Xi'an in Shaanxi. Again, in the early 600s, yet another capital was built at Xi'an. The forests of the Wei River valley and the Qinling Mountains, south of the city, took another huge hit. Edmonds writes, "Fuelwood needs in [Xi'an] were so urgent by the middle of the eighth century that the Tang rulers dug a canal to deliver cut timber from the southern mountain area to the city." Only the highest peaks of Qinling Mountains have remained uncut, probably to this day, due to their inaccessibility. Pandas still survive there.

The ninth through the thirteenth centuries also saw swells of deforestation. In each of these, the forests of Sichuan were exploited to make up shortfalls, and the migrant loggers contributed to the forest clearance. Finally, the Mongol dynasty moved its capital to Beijing from 1276 to 1367. While forests were cut around Beijing, forests around Xi'an recovered. The population crashes helped forest recovery in Sichuan. But for the most part, the rate of deforestation throughout China has increased and remained high during the last millennium, and demand continues to exceed supply.

In a poem titled "A Spring Night—Rejoicing in Rain," Du Fu describes a good rain:

> A good rain knows its season
> Comes forth in spring
> Follows the wind, steals into the night;
> Glossing nature, delicate without a sound.

This could almost be a prayer for those suffering the downslope effects of deforestation. The gentle shower is in marked contrast to the torrential rain and subsequent soil erosion, floods, landslides, and mud slips that are among deforestation's worst consequences. This relationship has been known in China for a long time. Edmonds quotes a Chinese text, the Guoyu, from between 608 B.C.E. and 590 B.C.E.: "Now if the mountain's forests are over exploited, the forests at the foot of the mountains will be destroyed, the swamps will be exhausted, the people's strength will be used up, and the fields will become devoid of crops and full of weeds . . . how will it be possible for there to be any happiness?"

Similar warnings were sounded during each crisis, and, if other events didn't intervene, actions such as reforestation continued to take place. The

Lakes clogged with logs headed for downstream mills are largely a thing of the past in giant panda habitat, thanks to recent changes in forestry policy.

Scenic attractions such as this lake in Jiuzhaigou Reserve draw thousands of tourists. Managing tourism in panda habitat is a future challenge for China's conservationists.

situation improved, but people forget. "Ecological amnesia" is the term for this forgetfulness, for the mental adjustments people make to change in their surroundings. Parents can tell their children how it once was, but the children will never truly comprehend what they haven't experienced.

In the early 1990s, social anthropologist Pam Leonard lived for a year in a western Sichuan mountain village called Xiakou, because she wanted to understand the "environmental consciousness" of farmers. The older villagers she lived among looked at the barren slopes around them and remembered the Old Society, the time before 1949, when their small village was set in thick forest and wild animals like leopards lived nearby, as if it had always been that way. Yet this area has likely been deforested several times in the past.

The flip side of ecological amnesia could be called "ecological nostalgia," hankering for a past of living in harmony with nature, although there is little evidence that people have ever lived in such a peaceable kingdom. As Leonard puts it, "Historical memory has two elements: what is remembered and what is forgotten." So, 2,600 years after the Guoyu's warning, Leonard wrote, "The soil, like the young people here, is engaged in a steady migration toward the towns below." Happiness remains elusive.

THE LONG MARCH

Historians as well as tour guides almost always say "Sichuan" and "refuge" in the same breath. Since the

ancient era of the Three Kingdoms, a time culturally equivalent to the Age of Chivalry in Western Europe, armies have retreated from the north and east over the mountains into Sichuan. In the 1930s, to escape the Nationalist forces of Chiang Kai-shek, Mao Zedong led his Communist troops on the epic Long March through Sichuan before regrouping in Shanxi, a grueling trek over cold mountain passes, across precarious bridges, and through dense forests—in fact, through places where the giant panda has also found refuge on the western edge of the Sichuan Basin. This is where the giant panda's most recent history has largely unfolded. The twentieth century in China was largely dominated by Chairman Mao, a man who said "to struggle against the earth is a constant joy."

Scholars agree that many of China's policies and programs under Mao, from 1949 to 1976, were environmentally disastrous. "Man Must Conquer Nature" was a constant refrain. Vaclav Smil first detailed the problem in his landmark 1983 book, *The Bad Earth,* and he and other scholars have continued to study the fallout. In her 2001 book, *Mao's War against Nature,* historian Judith Shapiro explains the damage done in so short a time by examining four themes.

One theme is political repression that stifled dissent. Mao was a proponent of population growth as a means to increase human capital. MaYinchu, one of China's most esteemed intellectuals and demographers, and once president of Beijing University, publicly disagreed in 1957 in a text titled *New Demography.* Ma predicted the problems that would arise from rapid growth and a large population, and urged the advocacy of family planning and incentives for small families. Not only was his scientific advice ignored, but he was also denounced and forced from his university position. In 1979, after Mao's death, the ninety-seven-year-old Ma was finally rehabilitated and *New Demography* has since become a classic. But the damage had been done. Shapiro cites Ma's biographer's estimate that the population of China would be 300 million people smaller if Ma's advice had been followed. That is, 900 million, not 1.2 billion.

A second theme is "utopian urgency." The revolutionary goal was "to remold the landscape quickly and achieve socialism." This led, among other excesses, to the Great Leap Forward, which began in 1958. Rural people were organized into large communes to increase agricultural production rapidly and overtake industrialized countries in steel production and other industries. Forests were cut indiscriminately to feed "backyard furnaces" to smelt fuel and, as Leonard describes, to feed the huge woks of the communal kitchens and the bonfires that burned during political education sessions that went long into the night. People were called from the fields to work at steel smelting, so crops were poorly tended.

People were also exhorted to "Wipe Out the Four Pests." Rats, sparrows, flies, and mosquitoes were the enemies. The campaign against sparrows, which eat grain, would be funny if it hadn't been a tragic success. Children were the front-line troops in this battle—along with smashing nests and breaking eggs, they made noise in the evening so the sparrows couldn't roost—but everyone pitched in. Shapiro writes, "Participants had to attack simultaneously or the sparrows would simply fly away to more tranquil places. But when millions of Chinese of all ages dispersed to the hillsides at the same hour to raise a ruckus, the sparrows had nowhere to alight." Ironically, sparrows also consume crop pests. No one knows how much grain was lost to the insects that flourished after the demise of their predators.

The overall result of the Great Leap Forward was famine and starvation that killed thirty to sixty million people between 1959 to 1962. The effects continued to reverberate into the future due to the loss of natural resources. One estimate cited by Shapiro suggests that 10 percent of Sichuan's forests were cut down in this short time alone.

The next phase in this history is captured in "Take Grain as the Key Link" to greater agricultural productivity. It exemplifies Shapiro's third theme, "dogmatic uniformity," and is often referred to as "learning from Dazhai," an exemplary commune held up as a model for others. Simply put, Chinese farmers were told to grow grain, and only grain, anywhere and everywhere. Wetlands were filled to grow grain. Forest was cleared to grow grain on entirely unsuitable steep slopes. The U.S. Department of Agriculture cautions against clearing

Construction scaffolding is among the thousands of products made of bamboo.

After bamboo flowers, it generally dies, leading to privation for animals and people dependent on it

Right: Removing trees that hold the soil on the steep slopes that characterize giant panda habitat leads to disastrous landslides and flooding.

moderate, 6 to 12 percent slopes; in China, slopes of 25 percent or greater were cleared and planted in grain. Age-old wisdom about which food plants did best in various soils and situations was ignored. In some places, even the fruit and other food and fiber trees most farmers grew were replaced with grain. Shapiro brings this home to our subject: "A scientist at the Chengdu panda breeding station grew up in Sichuan's Leibo County, where he recalled that as a child he often saw evidence of panda activity in the nearby hills, until 'learning from Dazhai' destroyed much of their habitat." This scientist commented, "I remember vividly how we cut bamboo and trees to build Dazhai-style terraces and grow grain wherever possible."

Finally came the "state-ordered relocations," Shapiro's fourth theme, especially those of the Cultural Revolution, when "educated youth" were sent to the countryside to work on such projects as creating irrigation systems and rubber tree plantations. Not only were some of these projects environmentally destructive in themselves, the young people had to eat and keep warm, and thus cleared forest to grow food and gather fuel.

Less well known were mass movements of laborers into remote regions of Sichuan to build factories and work in mines. This was the "Third Front" strategy, in preparation for war with both the United States and the Soviet Union. Emblematic is a huge steel mill that was built and put into production in Panzhihua, in southern Sichuan not far from Meigu. An influx of workers into the area also cleared land to farm; roads and a railroad were built through wilderness to move people and supplies; and air, land, and water pollution became severe. After 1965, when the project began, the mountains around Panzhihua were logged up to about 5,600 feet, and forest cover in the region fell to about 25 percent. Other mines and factories were dispersed through the region as well. Between Meigu and Xichang, which is in the same valley as Panzhihua, we drove through mile after mile of overgrazed alpine slopes. We saw almost no people and very few goats on the ground. Then, at 10,000 feet, we rounded a curve and confronted a factory, something to do with copper, we were told, right in what was once ideal giant panda habitat.

Mao Zedong died in 1976. In the same year, bamboo flowered and died over much of the giant panda's remaining habitat. It was the year of the dragon, when China expects change. It marked a turning point in China's political landscape and in the giant panda's. The first major panda census, taken at the end of 1975 and the beginning of 1976, revealed for the first time just how few pandas there were, and the bamboo flowering alerted the world to just how precarious their future was.

MOVING MOUNTAINS

Since the first economic reforms that followed Mao's death, China has pursued wealth with the same zeal that characterizes most of China's endeavors, and with much success. Twenty years separated our first trips to China from our second, casting change into stark relief and proving the saying that the past is another country. As has the United States, China has followed the "first you get rich, then you get clean" plan. Air and water pollution have become terrible problems in China. In a World Resources Institute ranking, nine of the ten most polluted cities in the world are in China. Farmland suffers from degradation; soil is lost to erosion. Deforestation, including logging of old-growth forest, continued until the ban of 1999, due in part to growing prosperity: Building a bigger house was one of the first things rural people wanted to do with their increased incomes. But there are many signs that China has reached the getting-clean stage.

Environmental concern is growing among the Chinese people. A 2000 survey of fifteen thousand Chinese from throughout the country found that 68 percent were willing to accept higher taxes for a cleaner environment. Most respondents believed the environment is one of the most serious issues facing the world and China, with 65 percent saying it is the world's biggest problem. Some 49 percent declared that environmental protection is China's greatest problem, followed by crime, overpopulation, and other issues. In one small sign of this concern, Shaanxi Province banned the production, sale, and use of disposable chopsticks. Statistics show that twenty-five million trees are cut down in China each year in

order to make forty-five billion pairs of disposable wooden chopsticks. And new programs are making the giant panda's future look less grim.

Many of the people Leonard knew in Xiakou lived through the environmentally calamitous Mao years, and their activities and attitudes toward the environment have been shaped by them. Before 1949, these villagers rarely cut down trees, and they planted a sapling for every tree they did cut. After, all the incentives encouraged exploitation. The years of unrestrained use and abuse of trees and other natural resources in common lands continued to manifest itself in behavior like tree stealing, even among neighbors. (Tree thievery has a long history in China, an indication of the value and scarcity of trees.)

Some people believe that someone else will get it—the tree or the herb—if they don't get it first. People remain uncertain about property rights that have only recently been given back to them and are thus less willing to invest in what might become communal land again. The rush to get rich has also changed people's relationship with their landscape. "Rising expectations of upward mobility and economic development can create an atmosphere in which people perceive their dependence on their own current resource base as temporary and changeable, which in turn makes them unwilling to invest in its long-term replenishment," writes Leonard.

Shapiro suggests that the problem of the commons might "have less to do with ownership than with lack of connection, responsibility, and good

More than a hundred species of bamboo grow in Sichuan. Some of the bigger ones are replacing wood as construction material and pulp for paper.

Right: Both people and pandas find fresh bamboo shoots appealing food. Sometimes, they compete directly for these nutritious morsels.

governance." And in China, good governance seems to go hand in hand with good natural resource management. Leonard points out that Chinese look to "strong and just leadership" to ensure that the natural competitiveness of people doesn't lead to environmental degradation. She told us that Chinese villagers would willingly refrain from activities harmful to giant pandas and their habitat if leaders explained openly and honestly what was expected of them, why it was necessary, and, most important, made everyone refrain equally.

BAMBOO DREAMS

In the Old Society, it was the custom on a spring evening for people to sit quietly in a grove of bamboo and listen to shoots burst—pop!—out of the sheathes that cocoon them. Imagine, listening to grass grow! Chinese also appreciate the susurration of the foliage in a breeze, calling it the "Sound from Heaven."

This tempts us to rush out and plant a bamboo grove in our garden, although noise in our urban soundscape might well suffocate the bamboo's song. And, honestly, we would probably regret planting a bamboo after it took over our small yard. For centuries in the Sichuan Basin, however, a patch of bamboo has stood near almost every rural homestead. Bamboo is prominent in all private

Arrow bamboo is a key panda food species in large areas of giant panda's range.

and public gardens, so much so that it represents all ornamental plants—hence, the opening lines of a poem by Du Fu:

> A famous garden lies near the green water,
> Numerous bamboos shoot up to the blue sky.

Bamboo is treasured for its aesthetic appeal, for the cool shade it creates, and for its symbolic value. It represents honesty and rectitude, and to live among bamboo is to enjoy a peaceful life. A Taoist saying goes, "A man can live without meat but not without bamboo." Just so, with a giant panda.

Bamboo is also immensely useful. Just as giant pandas depend on bamboo, so do countless numbers of people. In Asia, bamboo forms an extraordinarily important part of everyday life. Bamboo shoots and seeds provide food for people, and stems and leaves provide high-protein forage for livestock. People fashion bamboo into hats, baskets, mats, brooms, tools, toys, musical instruments, ornaments, objects of art, furniture, chopsticks, paper, fences, and weapons. Bamboo can be made into paper and plywood. With China's chronic shortage of soft wood, most of its paper is now made of bamboo pulp.

An important construction material, bamboo has twice the tensile strength of timber. An estimated billion or so people live in houses made of bamboo, which is cheap compared to wood. In Limón, Costa Rica, homes built of bamboo in the late 1980s in an effort to provide low-cost housing were the only ones standing after a 1992 earthquake. Workers scale bamboo ladders and scaffolding to build tall skyscrapers. These flimsy-looking structures are models of resilience, swaying in typhoons that might collapse steel frameworks. A bridge over the Min River in Sichuan hangs from cables of twisted bamboo—and has done so for more than one thousand years. All told, about 1,500 products—"from cradle to coffin"—are known to have been made of bamboo.

As useful a tool as bamboo is, cutting it requires a tool. Anthropologist Lynne Schepartz and her colleagues have been exploring a cave in southern China with a record of continued human occupation over the last 200 thousand years or more. In the cave are a surprising number of teeth of large animals, such as *Stegodon* and rhinoceros. Schepartz believes that people used these teeth as tools to cut bamboo because the stone in this area does not make good tools. It is impossible not to speculate about whether people used panda teeth in this way, too.

We tend to think of bamboo as typically Asian, and bamboos grown in North American gardens are largely species introduced from China and Japan. But many bamboo species grow and are extensively used in South America and Africa as well. The diversity of bamboos in South America rivals that of Asia, as does the diversity of its uses for people.

North America has only one native species. Called canebrakes, these towering bamboos were once common in the southeastern United States. They grew in huge expanses of dense stands that offered food and shelter to black bears, turkeys, and other wildlife. This giant bamboo had disappeared, though, by the 1950s, having been overgrazed by livestock and razed for agriculture. Scientists speculate that this may have contributed to the extinction of passenger pigeons and Carolina parakeets, and the decline of Bachman's warblers, which all ate the seeds and used the stands as feeding grounds and shelter.

Only a few wild animals besides pandas consistently eat the leaves and stems of bamboo. Shoots, however, are another matter. Apart from giant pandas, which Schaller suggested may eat between seventy and 100 pounds of shoots in a day, other shoot-eaters in panda country include takin, tufted deer, wild pigs, porcupines, badgers, hog badgers, palm civets, Asiatic black bears, and pikas. And why not? Shoots are succulent, high-protein morsels, and, in season, abundant.

People are fond of bamboo shoots, as well. To those accustomed to eating canned bamboo shoots, tasting fresh shoots—say, sauteed in chili and garlic as we ate them in Sichuan—is a revelation. They are buttery sweet and soft, like a barely cooked button mushroom, not at all like the fibrous slices we're used to. Cookbooks insist that fresh bamboo shoots must always be cooked to destroy the trace amounts of prussic acid, or cyanide, they contain.

SAVE THE SHOOTS

Soon after pandas Mei Xiang and Tian Tian arrived at the National Zoo, Susan received an e-mail from an anxious panda lover wondering whether she should boycott bamboo shoots. "Does harvesting bamboo shoots harm panda habitat?" she asked. The short answer turned out to be no—the shoots Americans buy in cans come largely or entirely from plantations.

But in the mountains of Sichuan, the story is different. In the spring, many people harvest wild bamboo shoots for personal consumption and local markets. This bamboo is of the same species that giant pandas crave. Yu Changqing, of the World Wildlife Fund, recently wrote, "When you visit panda habitat in spring, it is very difficult to see pandas or panda's traces, but it is very easy to meet people or see signs of their harvesting bamboo shoots. In fact, during the Panda Monitoring/Patrolling workshop [in 1999], bamboo shoot harvesting was identified as a threat to pandas. Recently in Qinling, the panda survey team met a serious case of bamboo harvesting in one place and the provincial government stopped the local people." Unfortunately, while denying people access to wild bamboo shoots may help giant pandas, it doesn't make the pandas popular with their human neighbors.

Many of the bamboos the people cultivate in the Sichuan Basin, and the ones generally grown commercially on plantations, are large species that don't seem to figure prominently in giant panda diets, although in the past, when pandas could reach them, they may have served as food during famines. Higher up into the mountains, however, we didn't see any of these backyard bamboos. Maybe it's too cold for these bamboos, or maybe in these agriculturally marginal areas, farmers don't spare the land to cultivate a plant that grows wild, because people here do use bamboo. We saw many piles of bamboo stalks, stacked next to houses, of the size generally preferred by giant pandas, but saw almost none growing outside of the reserves. Or at least none that we could see from the road or trail—giant pandas live outside of reserves so bamboo must be growing in less accessible areas. Among other things, bamboo is used for fences designed to keep animals out of fields. Harvesting bamboo for fences and other tools appears to be another arena of local competition between pandas and people, although perhaps one with the potential for a win-win outcome.

An organization called the International Network for Bamboo and Rattan (INBAR), based in Beijing, promotes bamboo from plantations as a substitute for timber from natural forests and for other uses such as papermaking and harvesting shoots. Not only would this industry save trees and wild bamboo, but it also makes economic sense because "bamboo is one of the fastest growing, shortest rotation and most productive forest resources in the world." More specifically, INBAR reports, "A 60-foot tree cut for market takes 60 years to replace. A 60-foot bamboo cut for market takes 59 days to replace."

INBAR encourages the development of bamboo growing and processing to improve local rural economies. There is a quick, three-to-five-year return on investment for a new bamboo plantation, and INBAR cites a study showing that income from plantation-grown bamboo shoots in one Chinese county moved many households out of poverty. This sustainable use of bamboo would make a sound contribution to human happiness, as well as to the conservation of giant pandas and their habitat.

Bamboo is plentiful all year around, making it a reliable food source—until, at long intervals, it flowers.

THE CONSERVATION CHALLENGE

"Prime habitat for giant pandas has been destroyed at an alarming rate within China's top nature reserve for the bamboo-munching creatures due to relentless human activity. . . ." This was the opening sentence in a widely disseminated news account in the spring of 2001. The story reported on a paper published in the prestigious journal *Science* titled "Ecological degradation in protected areas: The case of Wolong Nature

Previous pages: Even Wolong, China's flagship giant panda reserve, is not immune from the forces that are degrading the best panda habitat.

The panda's future depends on protecting the remaining habitat while also restoring connections between habitat fragments.

Reserve for giant pandas." The disturbing paper was written by Michigan State landscape ecologist Liu Jianguo (known to English speakers as Jack Liu) and his doctoral students, as well as scientists from the Chinese Academy of Sciences and the Wolong Nature Reserve. Their study found that although the 772-square-mile Wolong Reserve is thought to protect about 10 percent of the world's remaining giant pandas, "Local people living in the reserve have been the direct driving force behind the destruction of the forest and of giant panda habitat. . . ."

Wolong is a landscape of sharply incised valleys and lofty serrated slopes on the western escarpment that separates the flat Sichuan Basin and the Tibetan Plateau. It is just 81 miles from Chengdu, in the Qionglai Shan, one of the six mountain ranges where wild pandas still live. The headquarters is in the center of the reserve; its first home was a cluster of buildings left by the logging commune that was closed when Wolong became China's flagship giant panda reserve in 1975. Here, the Sichuan Forestry Department also maintains its giant panda breeding station and panda research laboratories.

Entering the reserve at nearly 4,000 feet in elevation, the road continues to climb, topping out on the Tibetan Plateau at nearly 14,000 feet. The reserve's highest peaks rise to 20,505 feet. To begin with, however, only the areas in the reserve between 7,400 and 9,000 feet in elevation and with 15 or less percent slope, or about half the total area, constituted highly suitable habitat for giant pandas. Highly suitable habitat has mixed conifer and deciduous broadleaf forest with arrow and umbrella bamboo. Liu and associates analyzed changes in forest cover over the last thirty-five years inside and outside of the reserve using satellite images of the habitats. They found that rates of high-quality habitat loss inside the reserve boundaries were lower before the giant panda reserve was established in 1975. After 1975, rates of fragmentation and loss of high-quality habitats dramatically increased in the reserve and now exceed or equal those rates outside the reserve. The best giant panda habitats were especially hard hit. Estimated giant panda numbers fell dramatically, from 145 in 1974 to 72 in 1986. Panda numbers are believed to have continued to slide since then. Wolong is spectacular, but this report did not paint a rosy picture for the giant panda's future there.

TUNNELING TO A CONNECTION IN THE QINLING SHAN

The loss of giant panda habitat reported from Wolong mirrors giant panda habitat loss in the other five mountain ranges where giant pandas still survive. From the mid-1970s to the mid-1980s, giant panda habitat overall was reduced by an estimated 50 percent. In the northernmost range, the Qinling Shan, just south of Xi'an in Shaanxi Province, a team of scientists from the World Wildlife Fund (WWF) and Beijing University conducted an analysis of giant panda habitat similar to that of Lui. Based

on estimates derived from censuses completed in 1998 and 1999, about 230 wild giant pandas live in the Qinling. Nine species of bamboo grow in these mountains, but two species, *Bashania fargesii* and *Fargesia spathacea*, are the key giant panda foods, making them essential elements for pandas' survival there. These bamboos flourish on the warm, wet, fertile south-facing slopes in various forest types above 5,000 feet. The transition to intensive agriculture lies below. Battered by northern winds, the north-facing slopes of these mountains are too dry and cold to support the essential bamboos, and no pandas live there. As in Wolong, roads, logging, and other activities chopped up the core giant panda habitat. With reforestation and policies that encourage more environmentally benign human activities, however, the major fragments could be reconnected.

In total, about 700 square miles of core giant panda habitat exists today, only 45 percent of which is now included in nature reserves. Habitat outside of reserves is threatened by commercial forestry, especially clear-cuts. Bamboo seedlings experience high mortality in clear-cuts, so bamboo fails to regenerate after it flowers and dies. Logging roads give access to loggers, but also provide avenues of encroachment for poachers and for people collecting firewood and other forest products such as herbs and mushrooms. Moreover, roads profoundly affect the behavior and movement of wildlife.

Yet splitting the core Qinling giant panda habitat is the heavily used National Road 108, which cuts across the top of the mountains to connect Xi'an and northern Sichuan. Fortunately, a major tunnel through the mountains is under construction to create a safer and more direct passage. This will be good for pandas. The high mountain road will fall into disuse and, without this barrier, the eastern and western blocks of the core panda habitat will again be connected.

There has been ongoing logging in old-growth spruce, fir, and pine forests to keep pace with China's need for softwoods, which are in very short supply. Most of the logging has occurred in the upper watersheds of the Yangzi and middle watersheds of the Yellow River. Thus, although overall forest cover has increased in China during the last half century to about 14 percent of the land area, logging in the remaining panda habitat was a continuing problem.

The hugely destructive flood of the summer of 1998 caused more than $20 billion worth of damage, affected 240 million people including more than 3,000 fatalities, and flooded 96,500 square miles of farm land. Previous economic gains from harvesting forests, including the precious softwoods, and from increased food production after forest clearance were offset by the devastating floods and landslides that also destroyed downstream transportation networks and hydropower facilities.

This epic inundation resulted in a sharp increase in China's awareness of the importance of protecting these vulnerable upper-watershed areas, which also contain panda habitat, to reduce flooding. Two-thirds of China's cultivated fields are in the flood-prone catchments of the Yangzi and Yellow Rivers. So the immense damage of the 1998 flood, and the likelihood of more floods in the future, produced a historic change in China's logging and land-use policy.

Strong measures to reduce the risk of flooding were declared a national priority, and, since 1998, a national decree banning commercial logging in these remote but vulnerable mountains has been in effect. The logging ban, later folded into the China Natural Forest Conservation Program, has been coupled with programs such as "Green-for-Grain" (Returning Steep Agriculture Slopes to Forest Program), which is restoring logged-over areas and reforesting erosion-prone hillsides of greater than 25 percent slope. These policies, carefully implemented, will turn tree cutters into tree planters and could make a profound difference in creating a secure future for wild giant pandas.

A tunnel, the logging ban, and the restoration of habitat; these seem a workable formula for sustaining giant pandas in the Qinling Shan in the short term. For the long term, protecting additional core habitat is crucial. Conservationists are also advocating forestry practices that promote

Wild giant pandas do not reproduce poorly, but they do reproduce slowly. This means they cannot quickly replace animals lost to poachers.

People harvesting wild food, such as ferns, in protected areas contribute to declining habitat quality.

Right: After devastating floods in 1998, China began to reforest steep slopes throughout the panda's range.

the growth of conifers; clear-cutting, for instance, promotes the growth of hardwoods rather than softwoods, and causes erosion and downstream flooding. Many conservationists also know that supporting and encouraging forest practices that involve and benefit local residents are key to long-term habitat and wildlife conservation.

These solutions are all part of an emerging, enlightened approach to securing a future for giant pandas and other large carnivores in the wild. Conservationists are increasingly talking about the 4-Cs formula: For large Carnivores to have a future, there must be Core protected areas in the form of reserves; habitat Corridors connecting the core areas; and the participation and support of human Communities that affect and are affected by the cores and corridors. In China, conservationists and government officials are working to establish trust among themselves and with the local people most affected by the day-to-day activities of giant pandas, because trust is a precursor to cooperation.

WHY DO PROTECTED AREAS SOMETIMES FAIL TO PROTECT?

People matter in conservation. A prevailing conservation model follows the dictum, "good fences make good neighbors." But this can lead to battles between preservationists on the one hand and spoilers on the other. There are no pristine areas in Asia, no places free of human influence. Giant panda habitat is no exception. People have been using these lands for millennia, and in many cases have the legal right to continue to do so. Fences are tools, not solutions.

After documenting the loss of giant panda habitat to human activities in the Wolong Nature Reserve, Liu Jianguo and his associates looked for root causes. Core giant panda areas, such as Wolong, are absolutely necessary as one leg in the chair that supports wild giant pandas. We usually think that designating areas as nature reserves and national parks will provide a large measure of protection and ensure the permanence of a wildland area. In Wolong, as in many areas called protected in China and around the world, this has not been the case. Unless we understand why, maintaining wildlands and wildlife in perpetuity will be a laudable but unattainable goal.

Because local people were the direct and driving force behind the destruction of panda habitat in Wolong, the scientists wanted to know how their attitudes, needs, and demography (ages and sex ratio) were affecting this giant panda protected area. Between 1975 and 1996, the number of people living inside the Wolong Nature Reserve increased by about 66 percent, to 4,336. The number of households increased by 150 percent. Each woman had, on average, two-and-a-half children. China's one child per couple rule does not apply to the minority ethnic groups—Tibetan, Chang, and Hui—who make up 70 percent of the people living in the reserve. The labor force (people twenty to fifty-nine years of age) in these communities increased by 45 percent

between 1982 and 1996, much faster than the total population.

Reserve residents are farmers who raise mostly corn and vegetables. They also collect firewood, harvest timber and bamboo, build wooden houses, and collect medicinal herbs from the forest. Some work for wages on road construction crews or at the panda research station; others are employed by the reserve's growing tourist industry. More boys than girls were born between 1982 and 1996. Eventually, these young men will marry young women from outside the reserve, and thus add people to the labor force. All of these people who must live and eat and work here are the root cause of giant panda habitat destruction, but could be part of the solution.

Conservation education activities have tended to focus on enhancing public awareness and participation. But does this kind of education encourage people to move out of the reserve to conserve panda habitat? Between 1982 and 1996, illiteracy among these people dropped from 31 percent to 25 percent. People living here truly value education. Liu and his colleagues found that young people would be willing to move out of the reserve to cities before they get married, especially if they could receive a higher education or more job opportunities. Older people preferred not to move, but took pride in their children and grandchildren going to college. Thus, emigration by young people here has strong family support.

In the past, people living in some reserves have either been moved out or relocated to areas within the reserve where they would have less impact. Both of these approaches were attempted in Wolong, but local people did not think they could subsist on what was being offered in place of their old homes and, understandably, did not move. In the end, the number of people living in the reserve has continued to grow.

Liu and his associates used computer simulations to see what might happen if, through improved education and incentives, young people left. They found that if only 22 percent of the young people relocated (through attending college, taking other jobs, and marrying out), the number of people living in the reserve would be reduced from 4,300 to 700 by the year 2047. Giant panda habitat would recover and then increase by 7 percent. Under the status quo, the reserve's human population would increase to 6,000 and panda habitat would be reduced another 40 percent by 2047.

The message to conservationists is that conservation education programs emphasizing awareness, protection, and participation are all well and good, but will not relieve the human pressures on giant pandas habitats. This will come only through training that readies young people for jobs in a rapidly changing world and working with the business sector to find suitable jobs for them. (China's entrance into the World Trade Organization may make this process inevitable, but painfully rapid.) There is a renewed strong emphasis in China on conflict resolution through mediation rather than compelling change by sheer force of will, which was the modus operandi under Mao. These results suggest a management option that can lead to improved conditions for the giant panda and for people. The eminent tropical biologist and innovative conservationist Daniel Janzen puts it this way: in biodiversity and ecosystem survival, "the priorities are humans and their happiness."

QIANFOSHAN NATURE RESERVE, "MOUNTAIN OF 1,000 BUDDHAS"

Qianfoshan was established as a nature reserve only in 1993. Relatively small at about 70 square miles, Qianfoshan is a mountaintop island of conifer, deciduous hardwood, and bamboo forest surrounded by a patchwork of farms whose fields blanket the hillsides. Amid the farms, however, are scattered stands of trees, the encouraging result of the recently implemented policy of converting land with slopes greater than 25 percent from crops to trees. Many of the newly planted trees are mulberries—on which silkworms are raised—and walnuts, so the plantations also generate income for the farmers. Qianfoshan faces serious pressures: it is small, surrounded by people, and has limited funds. About half of all wild giant pandas live in the Min Shan, and this is the southernmost panda population in the range. In the late 1980s, giant panda conservationist John MacKinnon, working with the landscape ecologist Robert De Wulf, analyzed satellite images to document the expansion of human settlement on

Left: Only a little more than half of the remaining panda habitat is in established reserves. But efforts to conserve this habitat are complicated by traditional grazing practices.

Hunting and trapping birds and other animals for food is a traditional source of meat for rural people.

Scientists don't know the extent to which a panda's normal activities are disrupted by people living and working nearby.

these slopes. They found that twenty-five years ago, a narrow corridor linked the Qianfoshan giant panda population with the north, but this is no longer true. They also noted that the forests of Qianfoshan extended much further down slope.

In May 2001, we drove from Chengdu to Bei Chuan Xian (North River County), which administers a portion of the Qianfoshan Reserve, and spent the night in a small hotel managed by the county forestry office. The hotel provides housing for guests and income for the forestry department. The next morning, we drove along a major road being modernized over the entire 42 miles to the Dun Shang protection station just outside the Qianfoshan Reserve boundaries. The area surrounding the reserve is home to about 20,000 people. From their mountain perch, the refuge's estimated fifteen to twenty-five giant pandas (according to a 1987 survey) could watch the traffic and the farmers weeding their plots of potatoes and greens.

The Qionglai Shan range, where Wolong is, lies to the south and west of Qianfoshan, and is separated from the Min Shan by the tremendous valley of the Min River, a major tributary of the Yangzi, so there is no way to connect these two panda reserves. The road that connects Bei Chuan with the Min River town of Maoxian

Offering young people opportunities to leave farming behind and find other work will reduce the pressures on panda habitat.

divides the giant panda population in the Min Shan in half. To the north live the giant pandas in what conservationists call the north Min Shan population, or population A; to the south is population B. There is no possibility of a tunnel connecting these panda populations as there is in the Qinling Shan, but a habitat corridor can be established. The small size of the Qianfoshan panda population is a concern because of the chance that some environmental catastrophe—fire or flood, for instance—could prove fatal. The potentially deleterious effects resulting from inbreeding could also hasten the panda's extinction here. Moreover, new road construction in the Min Shan range threatens to cut the wild giant panda population into four or five fragments, according to a study conducted by Beijing University.

Giant panda biologists Lu Zhi and Pan Wenshi, working in the laboratories of Stephen O'Brien at the National Cancer Institute, estimated genetic diversity in the various populations of giant pandas by analyzing blood and hair samples from wild individuals and from giant pandas of known origins living in breeding centers and zoos. They found no significant genetic difference among animals with origins in Qinling, Min, Qionglai, Liang, and Xiangling mountains. (No samples were available from the Daxiangling Shan.) But there are genetic signatures that revealed an individual's home mountain range. This indicates that the separation among different mountain populations is relatively recent, a matter of a few thousand years or less. The genetic diversity was similar to a noninbred population, but barely so. Levels of genetic diversity were highest among the pandas in the Qionglai range, followed by those in the Min range, followed by the Qinling mountain populations. Any further divisions in the Min Shan panda population would increase the risks of inbreeding and extinction.

It took two hours on a rough road to reach the Dun Shang protection station at the edge of Qianfoshan. The station looks no different than any of the small houses we passed along the way, and indeed, it was the home of the men who are park guards. This was the only indication that a nature reserve is nearby. Another few miles up the road, we entered the reserve, but there was no sign to tell us that we had done so. Boundaries have not been marked here. Then, another half mile farther, the road ended. A foot-trail climbed steeply up the mountain—and we did, too, walking through dense stands of bamboo until we reached 5,700 feet. At this elevation, pandas might be seen in September, according to the station guards who accompanied us.

We were there to see for ourselves how the Smithsonian National Zoological Park's donation could best be used to support wild giant conservation. Between 1991 and 1998, China established seventeen new giant panda reserves. Qianfoshan, like most of the others, needs just about everything to become truly functional. Boundaries must be posted so people know where they aren't supposed to be. Basic biological surveys are required to find out which species are present and in what numbers, followed by regular monitoring to track how species are faring. The guards who patrol the reserves need more tools than machetes; they need basics like binoculars and camping gear, and, sometimes, decent trails to patrol on. They also need to build additional protection stations within reserves.

We understand the human dynamic at work in Wolong; here the problems are somewhat different. How do conservationists respond to the human processes that are isolating mountaintop reserves and turning them into habitat islands? Qianfoshan was a "paper park" until the first protection station was built in 1996. The reserve's area is divided among and administered by three different counties. County officials need a coordinated administrative plan with shared goals that reflect the giant panda's ecological needs. Trained managers aware of the basic requirements of giant pandas are essential. Too, clear conservation objectives that take the giant panda's needs into account and agreed-upon criteria for reserve management are important. So, too, are accommodating human needs and desires. Each year thousands of pilgrims trek up the "1,000 Buddhas Peak." How can they continue this ritual with the least possible impact on pandas?

Because human welfare is a central issue in conservation, several payment programs that encourage economic activities which do not degrade habitat are widely discussed in conservation circles. Redeploying loggers as tree planters, for example, will help protect and re-establish forests on slopes in potential giant panda habitat. Other ideas in this "conservation by distraction" approach include creating nonfarm employment and paying for flood and erosion control programs that also benefit pandas. Sichuan farmers, for instance, are being offered grain subsidies to take steep slopes out of production. But how do you guarantee that subsidizing such activities will motivate more conservation? "First we eat, then we do everything else," M. F. K. Fisher observed. People will generally do what is in their best interest. If farmers receive more benefit from protecting a habitat than for clearing it for crops, for instance, they will do so. That is the challenge facing conservationists: how to keep these mountaintop reserves from being the end game; how to make them launching pads to a better future for wild giant pandas and for people.

Roads crisscross even the remotest parts of Sichuan, creating hazards for pandas and other wildlife and giving access to hunters and gatherers.

A DAM PLACE: YELE NATURE RESERVE

Yele is a young reserve, located in the Xiangling Shan. It supports the southernmost population of giant pandas, after Meigu. It was here that in 1929 the Roosevelt brothers shot a giant panda. Like Qianfoshan, Yele was established in 1993. Yele's headquarters is in Mianning, the county seat, and consists of a single small office in the county forestry building, itself a crumbling hulk. When we met with reserve staff here, however, we heard good news. A new reserve headquarters,

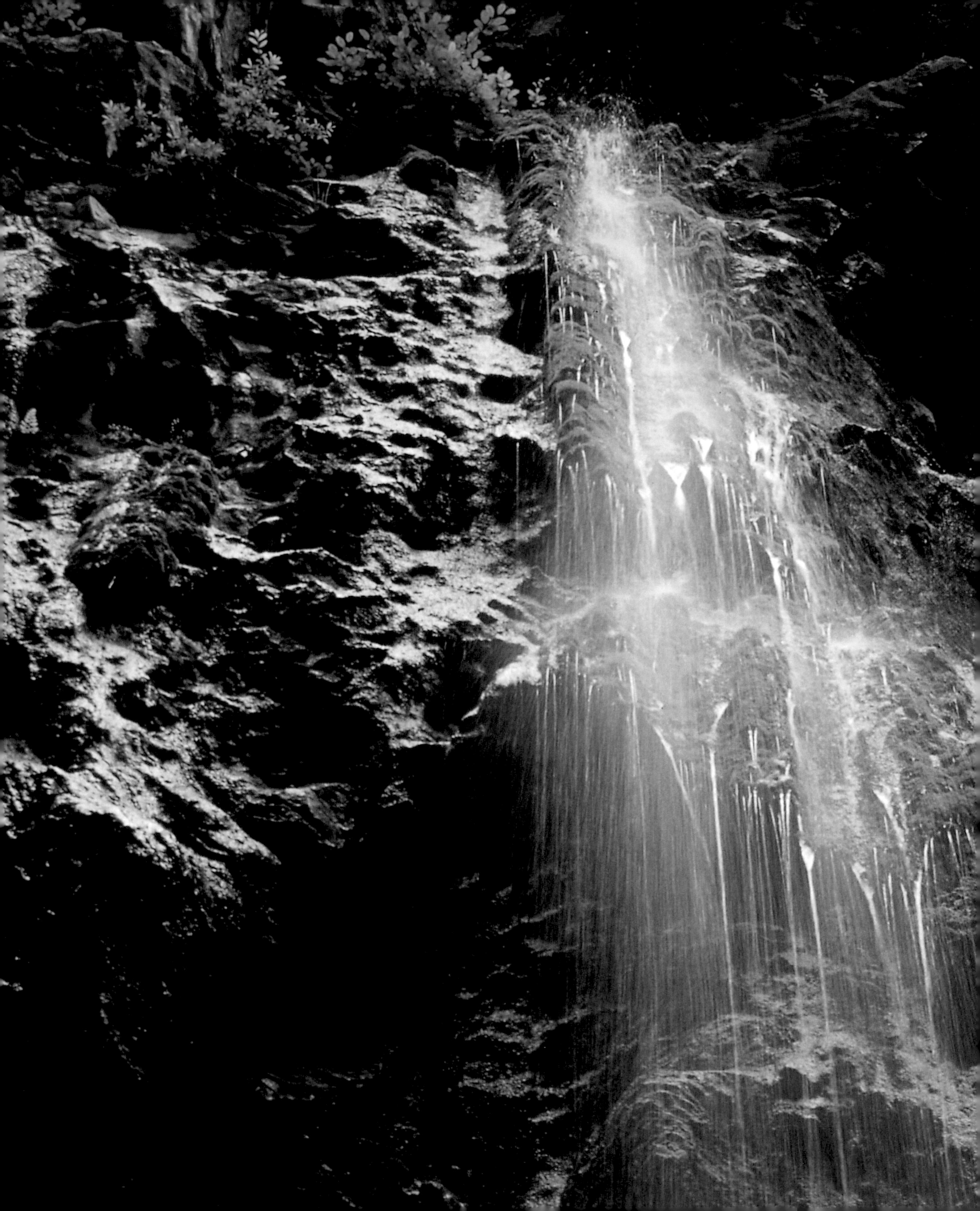

The spectacular beauty of giant panda habitat offers another compelling reason to conserve it.

which will include offices as well as an education center and staff housing, was nearing completion next door. Urban housing for reserve management staff, as well as the families of guards living in distant protection stations, makes it possible for reserve managers to attract and retain qualified staff. Like all of us, these people want access to housing that provides good schools for children, job opportunities for spouses, and proximity to medical care. But other than the headquarters and one protection station, this reserve also needs just about everything: from a basic biological survey and boundary stakes, to improved trails for patrolling, to communications and firefighting equipment.

It was rainy and cool when we left the hotel in Mianning for the reserve. The travel time was two-and-a-half hours, most of it on a paved, one-lane road that climbed along a steep, rocky slope on the right. On the left, broad farmlands stretched along the river below. Extensive reforestation, mostly in Chinese pine, was evident between small settlements and farms that are growing mostly corn and potatoes. At 7,500 feet, we noticed a little wild bamboo, a species that giant pandas probably eat. At this point, we also began to see the piles of wood and cut bamboo that seem to signal greater poverty among the farmers. That people were harvesting bamboo may account, in part, for the spotty distribution of bamboo in this area, and may indicate that people were competing with pandas for this resource.

After about two hours, we reached the muddy main street of a small

town on the Nan Ya (South Fork) River called Lizi Ping. This is the town nearest to the reserve, which is another thirty minutes' drive away, and where reserve staff have to go to make a phone call. Improving reserve communications is a necessity. In China, Yele's system is standard. Reserve headquarters are situated in the county town, often far from the reserve itself. In three reserves we visited, the minimum travel time between headquarters and the protected area itself is more than two hours. Lacking vehicles and telephones or radios, communications is, as the Yele staff put it, " by human relay"—pony express without the pony.

We turned off the main road onto a gravel one that continued to climb steeply. There were a few small villages along this road, but otherwise fewer people and more forest than anywhere else in Sichuan we visited. Then at 8,000 feet, we encountered a huge construction site, complete with earth-moving equipment, a shantytown of temporary housing for workers, and small mountains of rocks, sand, and logs standing in fields of mud. In a line of about a dozen tiny stores and teashops along the road, however, we saw Yele Nature Reserve's Three Forks River protection station. It stood out among the shop shacks because of the colorful mural of giant pandas and other wildlife painted on the station's streetfront. A little shocked, we learned that the construction, well inside the reserve boundaries, was for a dam and hydroelectric plant. This seven-year project will provide local power as well as power for export. In the long run, we realized later, this may benefit the reserve if local people no longer need to collect wood for fuel and building material. Right then, the scene was disheartening. So too was the evident logging around the site, although the explanation seemed reasonable that logging was approved only because this forest would eventually be inundated when the dam was complete.

The forest beyond the construction mess looked good, but there was road access for only another half-mile or so from the protection station, and the walking trails hadn't been maintained for years. The trails followed what appeared to be old logging roads, probably dating to the 1930s or before. The rain made what trails did exist dangerous. Nonetheless, some of our group set out on one of the two slippery paths along the river.

The beautiful valley was marred by a mine, the remains of bamboo that had been cut to make fencing for fields, and the conspicuous absence of large conifers. Cow and horse feces dotted the broken trail. But above the logging and the mining was magnificent forest with magnolia, at least four species of rhododendrons in flower, birch and other broadleaf deciduous trees, and very tall, mixed conifers, with a bamboo understory. This appeared to be as complete a forest assemblage as we had seen in any panda reserve. We were also encouraged to hear about plans to more than double the size of the Yele Reserve to about 230 square miles. Only perhaps eight pandas live in the reserve, so expanding the area of habitat to support more pandas seems wise.

After hiking for about an hour we turned back, cold, wet, and covered with mud. Our attempt to walk a short distance along the other path was thwarted by mudslides that closed the bit of road leading to the trailhead. A line of big blue trucks was stuck ahead of us, waiting. We were forced to back our vehicles up along the rutted muddy road that overlooked a river 100 feet straight below, until the road widened from the width of the jeep to its length, enabling us to make a delicate U-turn.

We left Yele a little frustrated, but pleased at the good habitat we'd seen. However, the hours there also gave us a chance to see how the guard staff lives. The electricity at the protection station was out that day, perhaps due to the rain, so there were no lights or heat in either the station office or the residences. Even if the power had been working, lighting would have been poor by U.S. standards. The one-room office, for instance, was lit by a bare bulb hanging from the ceiling and by whatever light came in the door (left open to the cool, damp air) and the small windows at either end. The floor was cold concrete. The furniture consisted of a narrow wooden bench lining three walls, a couple of small tables, and a few hard chairs. That was it. Yet maps of the reserve and colorful posters covered the walls, and staff pointed proudly to the spots on the map where pandas had been sighted in the last year.

A guard's residence, across a courtyard from the office, was no more

Pandas are rarely so easily seen in the wild. This makes it very difficult for conservationists to count them and determine how they are faring. Policies that encourage replanting pines and other conifers, which shade bamboo as well as plants like rhododendrons, will promote panda conservation.

Living conditions for nature reserve guards are precarious: this guard post was obliterated in a landslide soon after this photo was taken.

cozy. This typical concrete-floored apartment consisted of just two rooms: a front sitting/sleeping/working room that was about 12 feet square, and a rear kitchen that was about 5 feet by 12 feet. As in the office, a single bare bulb illuminated each room—or would if the electricity had worked. Light (and cold) came in the open door and through a window at the front and another at the back over the sink.

Furniture was spartan. A single wooden bed was covered with a few ragged but clean blankets, a quilt, and a pillow, but no mattress. A low, 3-foot square pine table draped with a red felt cloth stood between two backless, single-seat wooden benches. A small side table, about 3 feet long and 18 inches wide, held a couple of tea jars, a thermos, a flashlight, a small stack of brochures in Chinese explaining CITES regulations on wildlife trade, two other paperbound books, and a black-and-white framed photo of a young woman and child. This side table included a drawer, padlocked closed, but there was no sign of other possessions or of storage. (Under the bed, perhaps?)

The outhouse, which lacked running water and thus flushing ability, was around the corner; in the frequent rains, you would get wet on the way. Somewhere, we could only assume, there must be a shower facility. In short, far from luxury digs. This glimpse of life at a protection station that was *new* reminded us how little the reserves are asking for to help save giant pandas and their habitats, and how badly this help is needed.

Looking at these spare, modest quarters, we were reminded again that it is people who matter and will make the difference in the giant panda's future. How will China recruit, train, and sustain a cadre of people to lead and manage giant panda conservation efforts? "Policy is enunciated; it is realized in action. . . . The actual physical task of carrying out an organization's objectives falls to the person at the lowest level of the administrative hierarchy. . . . The fire is extinguished not by the fire chief or captain but by the team of firemen who play a hose on the blaze . . . ," Herbert Kaufman wrote in his classic *The Forest Ranger: A Study in Administrative Behavior.* The protection station guards are those people on the front line in giant panda conservation. In a large measure, the future of wild giant pandas will be determined by their values, beliefs, and actions. Those of us who wish wild pandas well must remember this and support these men and women every way that we can.

TANGJIAHE NATURE RESERVE

The Tangjiahe Giant Panda Reserve is a stunning example of the miracles that money, motivation, and trained, inspired staff and leaders can accomplish. When John visited here in 1981, Tangjiahe was a fledgling reserve like Yele and Qianfoshan. Now it stands as a management model for all the other reserves to emulate.

To get to Tangjiahe, we left the lower elevations of the Sichuan Basin and immediately headed into the mountains through a winding river canyon. Immense expanses of rice paddies were quickly replaced by terraced, irregularly shaped plots staircased into the hillsides. New trees occupied those spots too steep to farm effectively. The human population seemed less dense, but the land remained well farmed as we began to climb into the mountains approaching Tangjiahe along the Qing Chuan (Green River). Near to the reserve, we reached Gong Nong, a village of 600 to 800 people. Here the Sichuan Forestry Department and the Tangjiahe Reserve, with international funding, have been actively involved in community development projects to reduce pressure on the reserve. The reserve helped build a hydroelectric power plant so people would not have to collect wood for fuel. The reserve has assisted with planting walnut and chestnut trees, both to support the reforestation effort and provide a new cash crop. The reserve and local villagers are also cooperating in beekeeping—an ecologically friendly enterprise that earns about $30,000 a year in honey sales—and in growing medicinal plants.

The entrance to the reserve proclaimed its status. A huge sculpture of a giant panda welcomed guests, as did the people who live there and staff the well-tended protection station. The entrance fee was posted as 30 RMB (yuan), or nearly $4. About ten thousand people visit each year, and the reserve staff hope to increase that number in the future. A graphic noted that 301 lived here in 1978, until they were moved when the reserve was established. These people farmed to support the logging commune that once stood where reserve headquarters does today. Most of this reserve had been logged, but forest has returned. Unlike the continuing erosion of giant panda habitat by the people who have remained in Wolong, things here are improving because people were relocated.

The reserve headquarters complex is another 7½ miles farther into the park, at the end of the road. There are offices, a lecture hall, a natural history museum, staff residences, a pleasant guest house, a dining area, and even a dance hall featuring karaoke in the evening. A new education center was under construction, and the reserve director asked Lucy Spelman for assistance in developing its content and outreach programs.

Beyond the headquarters, the only access farther into the reserve was on foot along the ragged, sometimes treacherous remains of a logging road. Running along and above a river, this road was washed out in a great flood in 1991. In three long walks, we saw fairly abundant signs of wildlife, including takin feces; tracks of wild boar, hog badger, palm civet, and leopard cat; and actual sightings of a porcupine, a muntjac, and a tufted deer. Piles of chewed-on bamboo indicated porcupines were busy in the reserve. The staff also pointed out places, especially former apple and pear orchards, where one can see rhesus and golden monkeys and Asiatic black bears at some times of the year. We saw no sign of giant pandas, likely because they were at higher elevations in May. This reserve

Reserves and local people are joining forces to generate income for both. Bee keeping for honey (top), traditional arts and crafts (left), and sustainable harvest of bamboo, mushroom, and herbs (right), are among the new economic activities.

is connected with additional panda habitat in adjacent forests and other reserves to form the northern Min Shan giant panda population A, one of the largest that remains.

Recent surveys have shown that overall giant panda habitat in Qingchuan County has continued to slide, but Tangjiahe may be the most successful of all the giant panda reserves. Established in 1978 as 154 square miles in area, the reserve is slated to include all the remaining panda habitat in the county. To an outsider, expanding the reserve would seem ideal, but to the reserve administration, it is a worrisome burden. With the expansion will come the challenge of finding ways to make money to support additional anti-poaching patrols, training, housing, and salaries and benefits for the additional staff, and public relations and participatory programs to gain local support. Here and elsewhere, all of this falls to the reserves, which face chronic budget shortages and are left to find ways to make up for shortages on their own.

For a decade, government work units have been encouraged to become more financially self-sufficient in China. There is a national trend toward decentralization, with more local and provincial control of economic activity. Like many public entities in the United States, from universities to zoos, reserves in China are becoming income centers. In the last twenty-five years, the business of Tangjiahe has shifted from producing trees for timber to producing trees for tourists. Taking care of tourists raises funds, too, as does honey-making. What's important is that business does not harm the reserve. Beekeeping is not destructive, but tourism carries risks as well as rewards. Too many tourists, if poorly managed, can hurt the very values of the reserves that make them attractive to tourists in the first place.

Westerners are sometimes surprised when discussions about conservation abruptly become discussions about business; we tend to think that giant panda conservation should be above the financial fray. But at the end of the day, the question remains: who will pay the bills? Society understands that you get what you pay for. The conservation of wild giant panda is no exception.

A PANDA UNDER A PEAR TREE

Our encounter with the wild panda sitting under a pear tree occurred when our hosts in the Pingwu County Forest Department were notified of a potential panda problem and went to investigate. We were traveling with Ginette Hemley and Peter DeBrine of WWF as guests of Lu Zhi of the WWF–China Program and the Sichuan Forestry Department. Lu Zhi had been a student affiliated with the National Zoo a few years ago; now she is one of China's conservation leaders.

A meeting between the forest department officers, Lu Zhi, and the farmers was in progress when the rest of us gave up hopes of another glimpse of the panda and gathered in the farmyard. The farmers, small-statured men and women dressed in clothing that signaled their Tibetan origins, were explaining just how a panda came to be under their pear tree. The young female, as it turns out, had spent the night with thirty or so goats in the goat stall under the barn. It appeared that she had been herded into the stall inadvertently the evening before while mingling with the goats being put up for the night. In the course of the night, the farmers had heard a loud disturbance in the stall but had been reluctant to venture out into the night to investigate. It wasn't until morning that they opened the stall to release the goats. And to their surprise, the panda also emerged from the stall. The goats went off to pasture, but the panda remained behind to eat some pears. The farmers then found two dead goats and one badly injured kid in the stall.

The farmers had lined up the victims on a large boulder, as evidence of their loss, and negotiations were in progress. The goats had not been eaten. The injured kid had blood only on one front leg. There didn't seem to be any puncture wounds on one goat; the other had puncture wounds on its neck and back. The issues on the table were compensation for their lost livestock and what to do if the panda returned. It was decided that the farmers would be compensated and Lu Zhi would have her team of students from Beijing University fly today to Chengdu and drive to this farmstead to monitor the panda's activity should it return. (Even in these remote mountains, Lu Zhi made these arrangement on her cell phone.) The students and forestry officials would

Farmers tolerate, even enjoy, giant pandas living nearby and traveling through their land so long as they are fairly compensated for any damage a panda might cause.

Right: There are about two dozen separate panda populations in China. Most of them have fewer than fifty individuals, not enough to ensure their long-term survival.

wait to see if the panda returned, and if she did, would capture and move her to the Wanglang Nature Reserve. We learned later that she did return that night and killed nine more goats; she was captured, and released in the Wanglang Reserve. We don't know why she turned to goat killing; this is rare behavior among pandas. In the past, this young, dispersing female panda might have been "rescued" and taken to a breeding center, but now she might become a member of the core wild breeding population within the reserve.

Chinese wildlife officials are increasingly aware, based on our expanding understanding of panda behavior and ecology, of what wild pandas need to survive and how human activities disrupt natural panda dispersal corridors in the Min Shan. The new conservation-action paradigm also recognizes that incentive payments for local people are key to long-term success. Lu Zhi told us she had several records of young female pandas dispersing from their natal areas. The farm our pear-eating panda visited lies right in a panda dispersal pathway. Undoubtedly other pandas will come this way in the future. What will motivate these farmers to contact authorities again if they are not satisfied with the outcome this time? Just and rapid compensation for lost resources seems a minimal incentive, but one that is often lacking when conservationists are seeking common ground with local people. A future program may include incentive payments for farmers to maintain these habitat connections, just as we in the United States take tracts of land out of production and place them in long-term "soil banks" under Department of Agriculture programs. Increasingly, conservation planners are recognizing that stabilizing and sustaining what we call middle landscapes—the key habitat connections between the core reserves—is of primary importance to both local communities and the future of endangered wild giant pandas.

During our May 2001 reconnaissance of giant panda reserves, we rested for a day in Xichang, a lovely lakeside city in southern Sichuan. It offered a break for our hard-working drivers and an opportunity for us to catch up on our notes. In the afternoon, we quickly abandoned our laptops when the National Zoo's giant panda research and training coordinator Mable Lam rushed in and said, "Come see a giant panda!"

Amazed, we found the panda in a garage in a small city park, less than half of a mile away. This youngish adult female was in the care of the staff of Yele Nature Reserve. Villagers had found her a few weeks earlier about 6 miles from the reserve. She was near death, apparently from starvation, as she weighed only about 110 pounds. She was given veterinary care and seemed to be getting a little better, but was still the proverbial skin and bones. When her recovery was more complete, staff told us, they would move her, first to a larger enclosure in the park, then back to the reserve for eventual release.

Lucy Spelman, herself a veterinarian, consulted with the veterinarians

and caretakers to evaluate the panda's condition; she felt confident that the female would pull through but was concerned about the process of releasing her back into the wild, especially if she were to become overly comfortable with people during her stint in the park. These episodes dramatically point to the need to develop strategies for so-called rescued giant pandas. Conservationists must define the conditions that justify rescuing a panda. The official must decide whether an individual is really sick or at risk of starvation, or is just passing through some human habitation on its way to another patch of forest. Unlike the case of our panda under the pear tree, it is not always clear.

PANDAS AND PRIORITIES

Where did the pear-eating panda come from? When we saw her, she was on a hillside farm operated by Tibetans (and a leper colony, we were to learn) deep in the Min Mountains on the northern rim of the Sichuan Basin. One possibility is that she climbed over the ridge from the Tangjiahe Reserve in the next county. Or she might also have come from other forest areas in Pingwu County. A 1998 trial panda survey, organized by the Sichuan Forestry Department and WWF, found about 230 giant pandas living in 1,058 square miles of forest in Pingwu County, but only about 30 percent of them lived in the county's three panda reserves. The survey discovered that the remaining 70 percent lived in other forest tracts stretching along the high mountain slopes of the county. Compared to the results of surveys conducted in the 1980s, the number of pandas in the county seems to have remained about the same, but the animals appear to have moved to higher elevations and are more concentrated in the reserves because of disturbances related to logging. These reserves and forests are where China's national priorities and the giant panda's basic needs intersect in a way that can define the future for wild giant pandas.

Central China's foresters, once devoted to cutting down trees, are turning to planting trees. Bamboo will soon grow in re-forested areas.

The conservation of remaining forests for protection against downstream flooding has become more important than the consumption of the softwoods they provided. Pingwu County officials emphasized to us that the challenge the county now faced was how to find other employment for displaced loggers. Logging formerly employed about 3,000 people here and provided revenues for schools, health clinics, road building and maintenance, and other community needs. Aside from humanitarian issues, at least some of those people looking to put food on the table could be expected to turn to poaching in reserves and other forest areas. The county representatives were therefore much interested in the integrated conservation and development project that WWF is pioneering in Pingwu County. Part of that plan involves increasing ecotourism. As we left the briefing, we saw evidence of progress: a contingent of media people was leaving on what tourism promoters call a familiarization tour, which is designed to generate free publicity.

THE WAY TO WANGLANG

Leaving the farm in November 1999, our little caravan, with Lu Zhi in the lead, continued on the road that wound along the river toward the Wanglang Reserve. John had visited here in 1981 as a member of a Smithsonian Institution–China Association of Science and Technology group surveying areas for a possible giant panda field study. One member of his group, the distinguished mammalogist and conservation leader Wang Sung, told about his 1968 expedition to Wanglang to study the natural history of the giant panda. At that time, his team had to walk for four days along a foot trail to reach the reserve boundary. In 1981, John's group could drive right into the reserve in two-wheel-drive vans, traveling past picturesque villages and farms where people were plowing fields with teams of yaks. The good roads were part of the logging legacy in these mountains. Now, in 1999, the landscape remained similar to what John remembered. Again we were traveling up this broad valley with occasional towns surrounded by broad, gently sloping open fields. Here again were people plowing their fields with yaks.

But at least one thing was new on this trip. We stopped at Beima Township, a minority settlement, for refreshments, and found ourselves being

ushered into a cramped smoky kitchen with a central open fireplace. Here we were serenaded by young village women wearing colorful traditional clothing, complete with two white pheasant tail feathers placed upright in their hair. As they sang and chanted and cooked, the ends of these tail feathers jiggled constantly, creating a festive feeling, enhanced by the strong wine downed in numerous obligatory toasts. Then we were fed freshly prepared pancakes and bits of meat from a communal pot. This was all part of a pilot program that featured local talent performing traditional rituals to attract visitors, in the hope that such cultural entertainments would translate into a new income source for villagers. Their goal is not to attract foreign ecotourists, but to increase visitation from people living in Chinese urban centers such as Chengdu and beyond.

Above this valley are forested ridges that, according to the 1998 pilot giant panda survey, were either inhabited by giant pandas living at low densities or were potential habitat for giant pandas. Valley-bottom villages like Beima Township threaten to significantly disturb the forested ridge habitats, but it is encouraging that the 1998 survey found more pandas living here than did a similar survey conducted in 1986.

ENCOUNTERING A POACHER

The 124-square-mile Wanglang Reserve includes the very head of this long valley where the canyon walls start to pinch in, and clouds collect and inundate the area with rain and snow. We arrived at the reserve boundary in late afternoon and proceeded to the headquarters a few miles inside the reserve. Headquarters is home to a small staff of rangers, researchers, managers, and support staff; it also offered basic guest quarters. Since then, a new visitor center has been completed at the reserve entrance. But it didn't appear to have changed much since John had been here eighteen years earlier: just a handful of spartan buildings around a courtyard, plus a little modern technology.

A freestanding basketball hoop, minus a net, occupied a central place in the courtyard, not far from the satellite dish. The buildings were sparsely heated and dimly lit, and there was no indoor plumbing, but computers hummed in research rooms. Power comes from a microhydroelectric plant, made possible by the area's many rivers. This plant had been under construction in 1981. As compensation for difficult living conditions, however, the site is beautiful, and there are pandas here. In fact, the best bamboo habitat for giant pandas lies up and down the canyon a few miles from headquarters at the lower end of the reserve and in a bamboo forest that lies adjacent to, but outside, the reserve.

Travel in Wanglang is via old logging roads running along the main drainage, the Beima He (river). Just before dusk, we took a walk up the road looking for tracks or whatever we might meet. John noted that bamboo had regrown since his last visit.

Left: High-elevation forests are burned annually to maintain grazing pastures for yaks and other livestock. This eliminates places for pandas to live.

Forest fires are also a problem in panda habitat.

At least two of the bamboo species here had flowered in 1976, and when John traveled here five years later, many dead bamboo culms still littered the ground under large spruce and fir trees. But far from there being large homogenous bamboo stands as in other parts of the giant panda's range, the bamboo here is patchy and restricted to lower sheltered valleys and slopes, where the climate is cool and moist. Wanglang is located at the very edge of a microclimate belt that supports bamboo; just over the major ridges, where there are wide swings in temperature and low precipitation, bamboo is absent, and so are pandas. At the highest elevations in the reserve are imposing rock faces that reminded us of the Sawtooth Range in Idaho. These reaches neither support bamboo nor offer a permanent home for pandas. This means that less than half of this reserve consists of habitat with bamboo to sustain giant pandas.

Just up the road from the headquarters was an enclosure of several acres with 8-foot cement walls. This enclosure was constructed when the bamboo first started to flower. The thinking at the time was that giant pandas would starve when the bamboo flowered and died en masse. A primary bamboo species found here does flower at about ninety-year intervals and its culms then die. It takes several years for the bamboo to reestablish itself from seeds. To save giant pandas, authorities had planned to capture and house some of them here so they could supplement their food and create a giant panda breeding center. All this had proved to be unnecessary in the end, because this reserve contains several species of bamboo, which did not all flower and die at once. We now know there are at least twelve species of bamboo in Pingwu County. Giant pandas adjusted their movement to take advantage of whatever bamboo species was available and also tracked the seasonal and elevational changes in the availability of tender shoots and other edible parts of bamboo. Pandas, it turns out, are more flexible about the bamboo species they will eat than they'd been given credit for.

On the road we were walking along were carnivore scats that may have been a dhole's. We scanned for sign of Asiatic black bears, leopards, musk deer, and takin. There were once tigers in this area at low numbers, but they were extirpated by the 1950s or earlier. Churned mud revealed the presence of pigs, but we couldn't tell whether they were wild boar or domestic swine.

Susan was walking in front in the gathering dusk when she signaled and whispered that she'd seen something reddish about 50 feet ahead. The russet fur of a musk deer, perhaps? Then we saw a blur of blue clothing and heard thrashing in the underbrush. Arriving at the spot, we found a yoke stick with bloody, meat-filled burlap bags tied on each end; the meat was takin. What Susan had spotted was the leading edge of the yoke's bloody package poking out of the trees into the path. There was no sign of the undoubtedly terrified poacher, who had retreated up the hillside into the forest and deep evening shadows. Lu Zhi and a Chinese colleague kept vigil at the site while we four Americans raced back to headquarters to alert the reserve staff. Lu Zhi could tell that more than one person was involved from the back-and-forth whistling she heard as she waited. But the hunters did not return, and the reserve rangers' pursuit failed to unearth them. Staff guessed they merely slipped home to a nearby village—and a meatless meal.

Hunting and snaring hoofed mammals like takin and deer for food in these forests has a long tradition. Giant pandas are at times captured inadvertently in snares set for these food animals. But will the giant panda's long-term future be sustained by catching and prosecuting a few subsistence hunters? Killing a giant panda or a takin is already a serious crime in China, so only those truly in need are likely to risk the consequences of being caught. It seems far better to pursue a giant panda conservation strategy in partnerships with local people, a strategy that offers alternatives to poaching. And this is one of the goals of integrated conservation and development programs: to encourage and reward local guardianship of giant pandas and panda habitat through sustainable economic incentives.

At dinner that evening and breakfast the next morning, one small part of this effort was in evidence. The meal's mushrooms figure in a plan to develop a market for wild mushrooms that are sustainably harvested. The breakfast honey was farmed locally, another potentially marketable food product.

Already the watershed protection value of the forests has been recognized nationally as essential for the economic viability of China. Can this recognition of critical value be extended to giant pandas and other wildlife? The model program that Pingwu County is pursuing addresses this question. The key components are forest zoning; developing sustainable forestry and nontimber forest products such as mushrooms; creating conditions for ecotourism to flourish, including the preservation of old-town Pingwu; and managing protected areas with the input of local people. In all of these efforts, the need for watershed protection, the needs of giant pandas, and the very real needs of people are recognized, and the underlying economic well-being of Pingwu

Left: A healthy zoo population of giant pandas is an insurance policy against disaster. Conservation initiatives that address the needs of both pandas and people will ensure we never need to collect on it.

The future of people in remote Chinese villages is tied to the pandas that live practically in their backyards, as well as to the global village to which they now belong.

County is supported. This will make giant pandas stars in a process of ecological recovery and economic success. We came away feeling that this process was off and running in Pingwu County, where people are learning that sharing space, and even a few pears, with pandas might pay off.

A NEW DAY FOR PANDAS

At the close of our May 2001 visit, we met in Chengdu with officials of the Sichuan Forestry Department, the State Forestry Administration, and the China Wildlife Conservation Association (CWCA). We discussed how the Smithsonian National Zoological Park's donation towards giant panda conservation, managed through CWCA, can help support giant panda reserves. It was reassuring that the most senior member of the delegation was Wang Fuxing, who for many years headed the Sichuan Forest Department and now runs CWCA. He is known for getting things done. Wang Fuxing, like John, comes from a forestry background, and they shared thoughts on fire-control philosophy and infrastructure development in parks. We all agreed on the critical need to train and retain a dedicated staff, the key to any successful natural resource management or business effort. The Sichuan Forest Department has largely shifted from a log-producing agency to a tree-planting agency. This policy shift gave an air of optimism to the discussion about the National Zoo's donation to support some of China's priority panda conservation actions.

There have been no clear conservation objectives and criteria in panda reserve management. Until recently, there has been insufficient policy support, as well as little awareness of needs and poor capacity to undertake giant panda conservation activities. Sound scientific information upon which to base conservation decision-making has been lacking. These are contentious issues, but in identifying and addressing them, the problem of "saving the giant panda" is being quantified in ways that make crafting solutions possible.

In 1993, the State Council of China approved the National Panda Program and began to invest in an effort to save them. A revised giant panda conservation plan is due out at the end of 2002. China has enlisted many organizations to assist in implementing this plan, including WWF, the Zoological Society of San Diego, and the Smithsonian National Zoological Park. China and the West are communicating after a long silence. China has made a substantial commitment to the giant panda's future, a commitment symbolized by assigning so senior and experienced an administrator as Wang Fuxing to oversee the task.

Science, programs, and policies won't save pandas. People will. The road ahead is difficult. Still, there is a process in place. There is room for everyone to contribute. The giant panda will survive, if we let it.

Giant pandas will survive with bamboo forests, dens for their young, and freedom from persecution. It's up to all of us to make room for pandas.

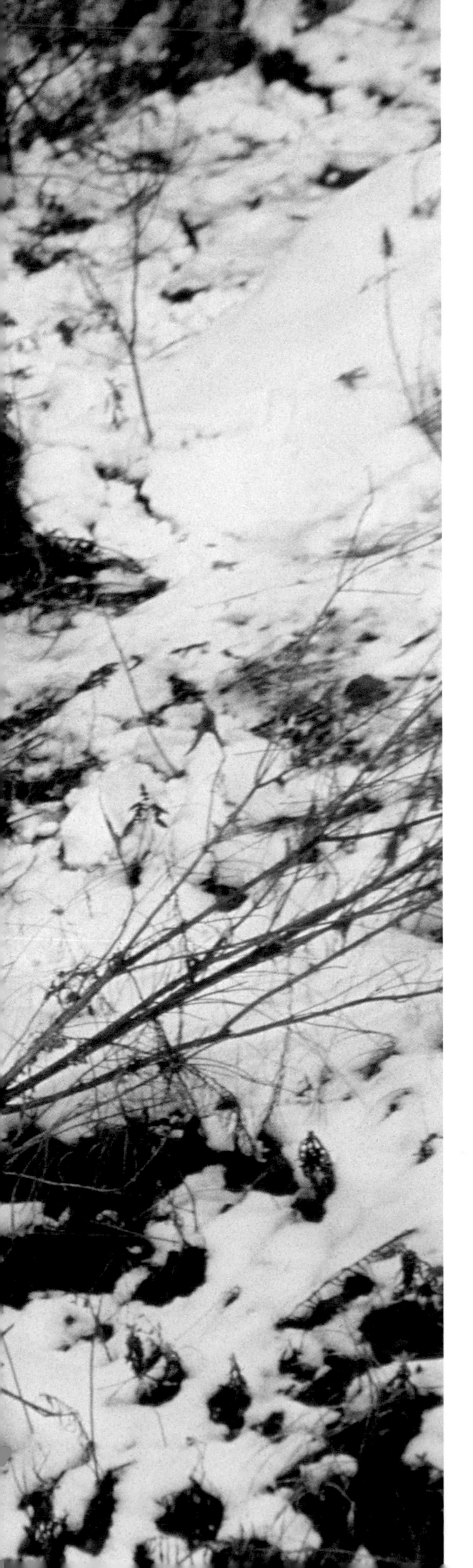

SCIENTIFIC NAMES OF SPECIES MENTIONED IN THE TEXT

Mammals

Diprotodontia	
Koala	*Phascolarctos cinereus*
Xenarthra	
Three-toed sloth	*Bradypus tridactylus*
Primates	
Golden bamboo lemur	*Hapalemur aureus*
Bamboo lemur	*Hapalemur griseus*
Greater bamboo lemur	*Hapalemur simus*
Macaque	*Macaca* species
Rhesus macaque	*Macaca mulatta*
Stump-tailed macaque	*Macaca thibetana*
Golden snub-nosed monkey	*Rhinopithecus roxellana* (sometimes seen as *Pygathrix roxellena)*
Leaf-eating monkey	*Presbytis* species
Gibbons	*Hylobates* species

Giant panda *Ailuropoda melanoleuca*

Mountain gorillas	*Gorilla beringei beringei*
Western gorilla	*Gorilla gorilla*
Human	*Homo sapiens*
Orangutan	*Pongo pygmaeus*
Carnivora	
Gray wolf	*Canis lupus*
Dhole (Asiatic wild dog)	*Cuon alpinus*
Fox	*Vulpes* species
Golden cat	*Catopuma temminckii*
Eurasian lynx	*Lynx lynx*
Bobcat	*L. rufus*
Pallas's cat	*Otocolobus manul*
Leopard cat	*Prionailurus bengalensis*
Clouded leopard	*Neofelis nebulosa*
Leopard	*Panthera pardus*
Tiger	*P. tigris*
Snow leopard	*Uncia uncia*
River otter	*Lutra lutra*
Badger	*Meles meles*
Hog badger	*Arctonyx collaris*
Yellow-throated marten	*Martes flavigula*
Least weasel	*M. nivalis*
Olingo	*Bassaricyon* species
Kinkajou	*Potos flavus*
Ringtail	*Bassariscus* species
Coati	*Nasua* species
Crab-eating raccoon	*Procyon cancrivorus*
Northern raccoon	*P. lotor*
Red panda	*Ailurus fulgens*
Giant panda	*Ailuropoda melanoleuca*
Sun bear	*Helarctos malayanus*
Sloth bear	*Melursus ursinus*
Spectacled bear	*Tremarctos ornatus*
American black bear	*Ursus americanus*
Asiatic black bear	*U. thibetanus*
Brown and grizzly bear	*U. arctos*
Polar bear	*U. maritimus*
Palm civet	*Paradoxurus hermaphroditus*
Masked palm civet	*Paguma larvata*
Order Proboscidea	
Asian elephant	*Elephas maximus*
Perissodactyla	
Horse	*Equus caballus*
Malayan tapir	*Tapirus indicus*
Sumatran rhinoceros	*Dicerorhinus sumatrensis*
Greater one-horned Asian rhinoceros	*Rhinoceros unicornis*
Javan rhinoceros	*R. sondaicus*
Artiodactyla	
Wild boar	*Sus scrofa*
Chinese forest musk deer	*Moschus berezovskii*
Siberian musk deer	*M. moschiferus*
Thorold's deer (white-lipped deer)	*Cervus albirostris*
Sambar	*C. unicolor*
Père David's deer (milou)	*Elaphurus davidianus*
Tufted deer	*Elaphodus cephalophus*
Muntjac	*Muntiacus* species
Roe deer	*Capreolus pygargus*
White-tailed deer	*Odocoileus virginianus*
Mule deer (black-tailed deer)	*O. hemionus*
Gaur	*Bos frontalis*
Yak	*B. grunniens*
Water buffalo	*Bubalus bubalis*
Takin	*Budorcas taxicolor*
Domestic goat	*Capra hircus*
Serow	*Naemorhedus sumatraensis*
Mountain goat	*Oreamnos americanus*
Musk ox	*Ovibos moschatus*
Bharal (blue sheep)	*Pseudois nayaur*
Chamois	*Rupicapra rupicapra*
Rodentia	
Fox squirrel	*Sciurus niger*
Chinese bamboo rat	*Rhizomys sinensis*
Porcupine	*Hystrix brachyura*
Lagomorpha	
Hare	*Lepus* species
Pika	*Ochotona* species

Birds

Order Galliformes	
Golden pheasant	*Chrysolophus pictus*
Lady Amherst's pheasant	*C. amherstiae*
White-eared pheasant	*Crossoptilon crossoptilon*
Reeve's long-tailed pheasant	*Syrmaticus reevesi*
Impeyan monal	*Lophophoris impejanus*
Chinese monal	*Lophophorus lhuysi*
Temminck's tragopan	*Tragopan temminckii*

Sichuan hill partridge	*Arborophila rufipectus*
Order Ciconiiformes	
Crested ibis	*Nipponia nippon*

Amphibians

Order Cryptobranchidae	
Appalachian hellbender	*Cryptobranchus alleganiensis*
Chinese giant salamander	*Andrias davidianus*

Plants

Trees	
Birch	*Betula*
Black locust	*Robinia pseudoacacia*
Cedar	*Cedrus*
Dawn redwood	*Metasequoia glyptostroboides*
Dogwood	*Cornus controversa*
Dove tree	*Davidia involucrata*
Eucalyptus	*Eucalyptus*
Fir	*Abies*
Larch	*Larix mastersiana*
Mulberry	*Morus*
Oak	*Quercus*
Chinese pine	*Pinus armandi*
Spruce	*Picea*
Walnut	*Juglans cathayensis*
Willow	*Salix*
Other Plants	
Astilbe	*Astible*
Camellia	*Camellia*
Cherokee rose	*Rosa laevigata*
Chinese ginseng	*Panax*
Daylily	*Hemerocallis*
Iris	*Iris*
Lily	*Lilium*
Peony	*Paeonia*
Rhododendron	*Rhododendron*
Wild rose	*Rosa*

Bamboo

The taxonomy, the formal naming, of bamboo species is in disarray for at least two reasons. You can tell bamboo apart with certainty and compare one closely related species with another only by examining their flowers. Scientists have to patiently watch a patch of bamboo for decades before they can obtain flowers. In taxonomy, scientists base their description of a species on the "type specimens" that are maintained in the world's large museums. Many of the bamboo type specimens were removed from China before and during World War II. When Chinese scientists began identifying bamboo species again after the war, there was no communication with Western museums, so they started over with new type specimens. The two systems have not yet been completely integrated. Thus, there is considerable confusion in the botanical literature as to what species are being referred to.

In addition, as scientists probe more deeply into giant panda habitat, they are finding what they believe to be many new species. These are tentatively named, but scientists must wait to establish with confidence their relationships with other bamboos until the new ones flower. The following names are the ones used by the scientists reporting on what the giant pandas they studied were eating:

Arrow	*Sinarundinaria*
Umbrella	*Fargesia*
Songu	*Fargesia spathacea*
Bashan muzhu	*Bashania fargesii*

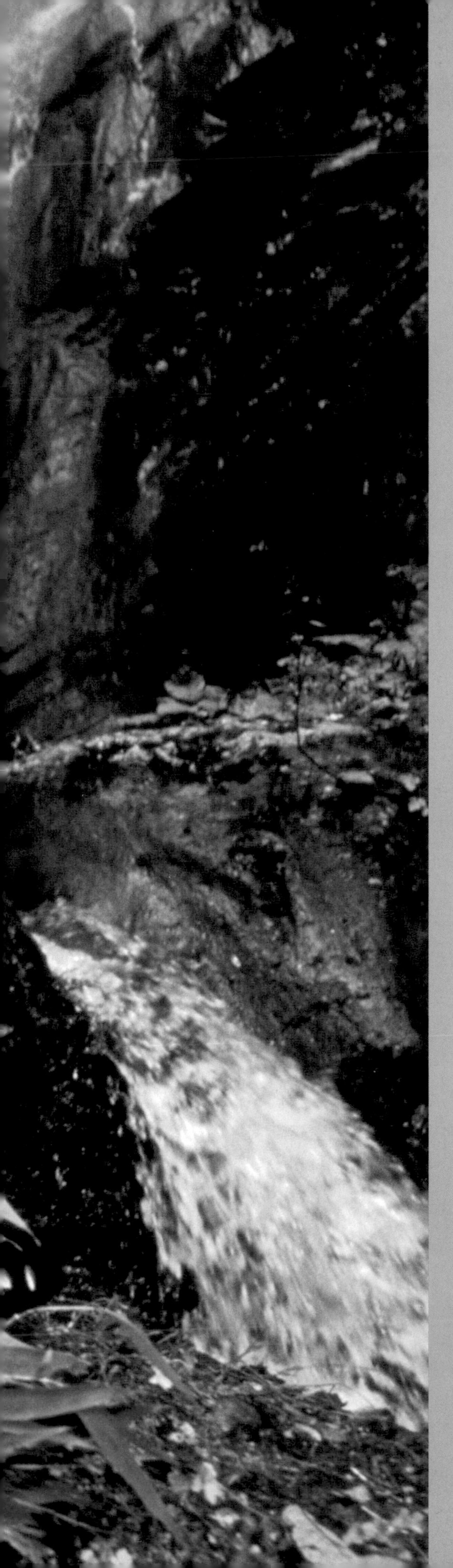

SOURCES AND SUGGESTED READING

PREFACE

2001. Studying giant pandas to save giant pandas: A special report to FONZ members. *ZooGoer* 30 (1): 8–13.

CHAPTER 1
PANDAS AND PEOPLE

The description of our meeting a wild giant panda under a pear tree first appeared in Seidensticker and Lumpkin (2001).

Kellert, S. R. 1996. *The Value of Life: Biological Diversity and Human Society.* Washington, D.C.: Island Press.

MacKinnon, J., and R. De Wulf. 1994. Designing protected areas for giant panda conservation. In *Mapping the Diversity of Nature,* R. I. Miller, ed., 127–142. London: Chapman and Hill.

Meffe, G. K., and C. R. Carroll. 1997. *Principles of Conservation Biology,* 2nd edition. Sunderland, Massachusetts: Sinauer Associates.

Schaller, G. B. 1993. *The Last Panda.* Chicago: University of Chicago Press.

Seidensticker, J., S. Christie, and P. Jackson, eds. 1999. *Riding the Tiger: Tiger Conservation in Human Dominated Landscapes.* New York: Cambridge University Press.

Seidensticker, J., and S. Lumpkin. 2001. Panda under a pear tree. *ZooGoer* 30 (1): 18–23.

CHAPTER 2
ICONS AND AMBASSADORS

Catton, C. 1990. *Pandas.* New York: Facts on File, Inc.

Gladwell, M. 2000. *The Tipping Point: How Little Things Can Make a Big Difference.* Boston: Little, Brown and Co.

Gould, S. J. 1992. *The Panda's Thumb: More Reflections in Natural History.* New York: W. W. Norton & Co.

Ignatius, D. 1999. The panda puzzle. *The Washington Post,* December 14, A38.

Lyon, E. 1992. "P-Day" at the National Zoo. *ZooGoer* 21 (2): 23–25.

McCafferty, D. 2001. Saving the world in glorious black and white. *USA Weekend,* June 23–24.

Morris, R., and D. Morris (revised by J. Baarzdo). 1981. *The Giant Panda.* New York: Penguin Books.

O'Connor, N. K. 1992. A quest for pandas in Chinese art. *ZooGoer* 21 (2): 19–20.

Robbins, S. 1976. Pandas and man. *ZooGoer* 4 (6): 19–23.

Roosevelt, T., and K. Roosevelt. 1929.

Previous pages: Millions of words have been written about giant pandas in the last 100 years.

An understanding of forest ecology and the behavior of trees illuminates the ecology and behavior of giant pandas.

Trailing the Giant Panda. New York: Charles Scribner's Sons.

Schaller, G. B. 1993. *The Last Panda.* Chicago: University of Chicago Press.

Seidensticker, J., J. F. Eisenberg, and R. Simons. 1984. The Tangjiahe,Wanglang, and Fengtongzhai giant panda reserves and biological conservation in the People's Republic of China. *Biological Conservation* 28:217–251.

CHAPTER 3
IN DEEP TIME

Bakken, D. A. 1997. Taphonomic patterns of Pleistocene hominid sites in China. *Indo-Pacific Prehistory Association Bulletin* 16:13–26.

Baskin, J. A. Procyonidae. In *Evolution of Tertiary Mammals of North America,* C. M. Janis, K. M. Scott, and L. L. Jacobs, eds., 144–151. New York: Cambridge University Press.

Bininda-Emonds, O. R. P., J. L. Gittleman, and A. Purvis. 1999. Building large trees by combining phylogenetic information: A complete phylogeny of the extant Carnivora (Mammalia). *Biological Review* 1999:143–175.

Chaimanee, Y., and J. J. Jaeger. Pleistocene mammals of Thailand and their use in the reconstruction of the paleoenvironments of Southeast Asia. *SPAFA Journal* 3 (2): 4–10.

Chinh, H. X. 1991. Faunal and cultural changes from Pleistocene to Holocene in Viet Nam. *Indo-Pacific Prehistory Association Bulletin* 10:74–78.

Ciochon, R. L. 1991. The ape that was. *Natural History* 100 (11): 54–63.

Ciochon, R. L., and J. W. Olsen. 1991. Paleoanthropological and archaeological discoveries from Lang Trang caves: A new Middle Pleistocene hominid site from northern Viet Nam. *Indo-Pacific Prehistory Association Bulletin* 10:59–73.

Davis, D. D. 1964. The giant panda, a morphological study of evolutionary mechanisms. *Fieldiana* 3:1–339.

Eisenberg, J. F. 1981. *The Mammalian Radiations.* Chicago: University of Chicago Press.

———. 1989. An introduction to the Carnivora. In *Carnivore Behavior, Ecology, and Evolution,* vol. 1, J. L. Gittleman, ed., 1–9. Ithaca, N.Y.: Cornell University Press.

Eisenberg, J. F., and J. Seidensticker. 1976. Ungulates in southern Asia: A consideration of biomass estimates for selected habitats. *Biological Conservation* 10:293–308.

Flynn, J. J., and M. A. Nedbal. 1998. Phylogeny of the Carnivora (Mammalia): Congruency vs. incompatibility among multiple data sets. *Molecular and Phylogenetic Evolution* 9:414–426.

Hunt Jr., R. M. 1996. Biogeography of the order Carnivora. In *Carnivore Behavior, Ecology, and Evolution,* vol. 2, J. L. Gittleman, ed., 485–494. Ithaca, N.Y.: Cornell University Press.

Hunt, Jr., R. M. 1998. Ursidae. In *Evolution of Tertiary Mammals of North America,* C. M. Janis, K. M. Scott, and L. L. Jacobs, eds., 174–195. New York: Cambridge University Press.

Kingdon, J.1993. *Self-made Man.* New York: John Wiley.

Mayr, E. 1982. *The Growth of Biological Thought.* Cambridge, Mass.: Harvard University Press.

Long, V. T., J. de Vos, and R. L. Ciochon. 1996. The fossil mammal fauna of the Lang Trang Caves, Vietnam, compared with the Southeast Asian fossil and recent mammal faunas: The geographical implications. *Indo-Pacific Prehistory Association Bulletin* 14:101–109.

Mugaas, J. N., and J. Seidensticker. 1993. Geographic variation of lean body mass and a model of its effect on the capacity

of the raccoon to fatten and fast. *Bulletin of the Florida Museum of Natural History* 36:85–107.
Mugaas, J. N., J. Seidensticker, and K. P. Mahlke-Johnson. 1993. Metabolic adaptation to climate and distribution of the raccoon *Procyon lotor* and other Procyonidae. *Smithsonian Contributions to Zoology* 542:1–34.
O'Brien, S. J., W. G. Nash, D. E. Wildt, M. E. Bush, and R. E. Benveniste. A molecular solution to the riddle of the giant panda's phylogeny. *Nature* 317:140–144.
Palomares, F., and T. M. Caro. 1999. Interspecific killing among mammalian carnivores. *American Naturalist* 153:492–508.
Schepartz, L. A., S. Miller-Antonio, and D. B. Bakken. 2000. Early palaeolithic occupation of southwestern China and adjacent areas of Vietnam and Thailand. *Acta Anthropologica Sinica* 19 (Supplement): 126–131.
Van Valkenburgh, B. 1999. Major pattern in the history of carnivorous mammals. *Annual Review of Earth and Planetary Science* 27:463–493.
Wayne, R. K., R. E. Benvesiste, D. N. Janczewski, and S. J. O'Brien. 1989. Molecular and biochemical evolution in the Carnivora. In *Carnivore Behavior, Ecology, and Evolution,* vol. 1, J. L. Gittleman, ed., 465–494. Ithaca, N.Y.: Cornell University Press.
Werdelin, L. 1996. Carnivoran ecomorphology: a phylogenetic perspective. In *Carnivore Behavior, Ecology, and Evolution,* vol. 2, J. L. Gittleman, ed., 582–624. Ithaca, N.Y.: Cornell University Press.
Xue, X., and Y. Zhang. 1991. Quaternary mammalian fossils and fossil human beings in China. In *The Quaternary of China,* Z. Zhang, and S. Shao, eds., 307–374. Beijing: China Ocean Press.

CHAPTER 4
GIANT GRASS-EATING BEARS

Campbell, J. J. N., and Z. S. Qin. 1983. Interaction of giant pandas, bamboo, and people. *Journal of the American Bamboo Society* 4:1–35.
Davis, D. D. 1964. The giant panda, a morphological study of evolutionary mechanisms. *Fieldiana* 3:1–339.
Gittleman, J. L. 1989. Lactation energetics and a general model of protracted growth in the red panda, *Ailurus fulgens.* In *Red Panda Biology,* A. R. Glatston, ed., 79–94. The Hague: SPB Academic Publishing.
Gittleman, J. L. 1994. Are the pandas successful specialists or evolutionary failures? *BioScience* 4:456–464.
Glander, K. E., P. C. Wright, D. S. Seigler, V. Randrianasolo, and B. Randriansolo. 1989. Consumption of cyanogenic bamboo by a newly discovered species of bamboo lemur. *American Journal of Primatology* 19:119–124.
Janzen, D. 1976. Why bamboos wait so long to flower. *Annual Review of Ecology and Systematics* 7:347–391.
Kleiman, D. G. 1983. Ethology and reproduction of captive giant pandas *(Ailuropoda melanoleuca). Zeitschrift Tierpsychologie* 62:1–46.
Lu, Z., and W. Pan. 1998. Genetic diversity of the giant panda population. In *International Workshop on Feasibility of Giant Panda Reintroduction,* S. Mainka and Z. Lu. eds., 233. Beijing: China Forestry Press.
Lu, Z., W. Pan, X. Zhu, D. Wang, and H. Wang. 2000. What has the panda taught us? In *Priorities for the Conservation of Mammalian Diversity: Has the Panda Had Its Day?* A. Entwistle and N. Dunstone, eds., 325–334. New York: Cambridge University Press.

Mahaney, W. C., S. Aufreiter, and R. G. V. Hancock. 1995. Mountain gorilla geophagy: A possible seasonal behavior for dealing with the effects of dietary changes. *International Journal of Primatology* 16:475–488.
McNab, B. K. 1989. Energy expenditure in the red panda. In *Red Panda Biology,* A. R. Glatston, ed., 73–78. The Hague: SPB Academic Publishing.
McClure, F. 1943. Bamboo as food. *Journal of Mammalogy* 24:267–268.
Pan, W. 1998. Ecology and behavior of the giant panda in the Qinling, with comments on captive panda reintroduction. In *International Workshop on Feasibility of Giant Panda Reintroduction,* S. Mainka and Z. Lu. eds., 224–227. Beijing: China Forestry Press.
Peters, G. 1985. A comparative survey of vocalizations in the giant panda (*Ailuropoda melanoleuca,* David, 1869). *Bongo* 10:197–208.
Schaller, G. B., Q. Tang, K. G. Johnson, X. Wang, H. Shen, and J. Hu. 1989. The feeding ecology of giant pandas and Asiatic black bears in the Tangjiahe Reserve, China. In *Carnivore Behavior, Ecology, and Evolution,* vol. 1, J. L. Gittleman, ed., 212–241. Ithaca, N.Y.: Cornell University Press.

Schaller, G. B., J. Hu, W. Pan, and J. Zhu. 1985. *The Giant Pandas of Wolong.* Chicago: University of Chicago Press.

Schmidt-Nielson, K. 1984. *Scaling: Why Is Animal Size So Important?* New York: Cambridge University Press.

Seidensticker, J., J. F. Eisenberg, and R. Simons. 1984. The Tangjiahe, Wanglang, and Fengtongzhai giant panda reserves and biological conservation in the People's Republic of China. *Biological Conservation* 28:217–251.

Stirling, I. 1988. *Polar Bears.* Ann Arbor: University of Michigan Press.

Swaisgood, R. R., D. G. Lindburg, and X. Zhou. 1999. Giant pandas discriminate individual difference in conspecific scent. *Animal Behaviour* 57:1045–1053.

Swaisgood, R. R., D. G. Lindburg, X. Zhou, and M. A. Owen. 2000. The effects of sex, reproductive condition, and context on discriminations of conspecific odours by giant pandas. *Animal Behaviour* 60:227–237.

Taylor, A. H., D. G. Reid, Z. Qin, and J. Hu. 1991. Bamboo die-off: An opportunity to restore panda habitat. *Environmental Conservation* 19:76–79.

Taylor, A. H., and Z. Qin. 1997. The dynamics of temperate bamboo forests and panda conservation in China. In *The Bamboos,* G. P. Chapman, ed., 189–203. London: Academic Press.

Zhu, X., D. G. Lindburg, W. Pan, K. A. Forney, and D. Wang. 2001. The reproductive strategy of giant pandas *(Ailuropoda melanoleuca):* Infant growth and development and mother-infant relationships. *Journal of Zoology, London* 253:141–155.

CHAPTER 5
THE PANDA'S UMBRELLA

Parts of this chapter first appeared in Lumpkin and Seidensticker (2001).

Boufford, D. E., and S.A. Spongberg. 1983. Eastern Asian–eastern North American phytogeographical relationships: A history from the time of Linnaeus to the twentieth century. *Annals of the Missouri Botanical Garden* 70:423–439.

Camp, W. H., V. R. Boswell, and J. R. Magness. 1957. *The World in Your Garden.* Washington, D.C.: National Geographic Society.

Chen Changdu, et al., eds. 1998. *China Biodiversity: A Country Study.* Beijing: China Environmental Science Press. (SEPA, ISBN 7-80135-256-4).

Dowell, S. D., B. Dai, R. P. Martins, and R. S. R. Dowell. 1998. The Sichuan hill partridge forest conservation project. *Species* 30:13–14.

Fox, H. M. 1949. *Abbé David's Diary.* Cambridge, Mass.: Harvard University Press.

International Network for Bamboo and Rattan at http://www.inbar.int/ includes links to articles and information on bamboo, its uses, and its cultivation as a sustainable crop.

Johnson, K. G., W. Wang, D. G. Reid, and J. Hu. 1993. Food habits of Asiatic leopards *(Panthera pardus fusea)* in Wolong Reserve, China. *Journal of Mammalogy* 74 (3): 646–650.

Judziewicz, E. J., L. G. Clark, X. Londono, and M. J. Stern. 1999. *American Bamboos.* Washington, D.C.: Smithsonian Institution Press.

Kirkpatrick, R. C. 1995. The natural history and conservation of the snub-nosed monkeys (genus *Rhinopithecus*). *Biological Conservation* 72:363–369.

Lumpkin, S., and J. Seidensticker. 2001. Under the giant panda's umbrella. *ZooGoer* 30 (1): 24–29.

Mittermeier, R. A., N. Myers, and C. G. Mittermeier. 2000. *Hotspots: Earth's Biologically Richest and Most Endangered Terrestrial Ecoregions.* Chicago: University of Chicago Press.

Neas, J. F., and R. S. Hoffmann. 1987. *Budorcas taxicolor. Mammalian Species* 277:1–7.

Olson, D. M., and E. Dinerstein. 1998. *WWF's Global 200: A Representation Approach to Conserving the Earth's Distinctive Ecoregions.* Document at http://www.wwfus.org/news/pubs/g200 pdf.pdf.

Schaller, G. B., Q. Teng, W. Pan, Z. Qin, X. Wang, J. Hu, and H. Shen. 1986. Feeding behavior of Sichuan takin *(Budorcas taxicolor). Mammalia* 50:311–322.

Sheng, H., O. Noriyuki, and H. Lu. 1999. *The Mammalian of China*. Beijing: China Forestry Publishing House.

Spongberg, S. A. 1993. Exploration and introduction of ornamental and landscape plants from eastern Asia. In *New Crops*, J. Janick and J. E. Simons, eds., 140–147. New York: Wiley.

UNEP and WCMC. 2001. Crested ibis *Nipponia nippon*. At http://www.unep-wcmc.org/index.html?http://www.unep-wcmc.org/species/data/species_sheets/crestedi.htm~main\

Wan, B., and C. Liu, project directors. 1994. *China: Biodiversity Conservation Action Plan*. Beijing: National Environmental Protection Agency.

Wen, J. 1999. Evolution of eastern Asian and eastern North American disjunct distributions in flowering plants. *Annual Review of Ecology and Systematics* 30:421–455.

Wilson, E. H. 1913, Methuen & Co. Ltd.; reprint ed., 1986. *A Naturalist in Western China*. London: Cadogan Book Ltd.

CHAPTER 6
CHINESE SHADOWS

Bird, I. L. 1899; reprint ed., 1987. *The Yangtze Valley and Beyond: An Account of Journeys in China, Chiefly in the Province of Sze Chuan and among the Man-tze of the Somo Territory*. Boston: Beacon Press.

Edmonds, R. L. 1994. *Patterns of China's Lost Harmony*. London: Routledge.

Elvin, M. 2001. The Retreat of the Elephants. Unpublished manuscript.

Glahn, R. von. 1987. *The Country of Streams and Grottoes: Expansion, Settlement, and the Civilizing of the Sichuan Frontier in Song Times*. Cambridge, Mass.: Harvard University Press.

Hartwell, R.M. 1994. Fifteen centuries of Chinese environmental history: Creating a retroactive decision-support system. Document at http://citas.csde.washington.edu/org/report_c.

Ho, P.-T. 1955. The introduction of American food plants into China. *American Anthropologist* 57:191–201.

Leonard, P. 1994. *The Political Landscape of a Sichuan Village*. Unpublished dissertation, Department of Social Anthropology, Darwin College, University of Cambridge.

Lumpkin, S., J. Seidensticker, and L. Spelman. 2001. Around the rim in fourteen days: Travels into four giant panda reserves in the mountains that rim the Sichuan Basin, May 2001. Document at http://pandas.si.edu/reserve/introduction.htm.

Marks, R. B. 1998. *Tigers, Rice, Silk, and Silt: Environment and Economy in Late Imperial South China*. Cambridge: Cambridge University Press.

Pomeranz, K. 2000. More "Malthusian Mythologies"? Rethinking living standards, environment, and "population pressure" in the 19th-century lower Yangzi. Unpublished manuscript.

McNeill, W. H. 1991. American food crops in the Old World. In *Seeds of Changes*, J. Viola and C. Margolis, eds., 43–59. Washington, D.C.: Smithsonian Institution Press.

Shapiro, J. 2001. *Mao's War against Nature: Politics and the Environment in Revolutionary China*. Cambridge: Cambridge University Press.

Smil, V. 1983. *The Bad Earth*. Armonk, N.Y.: M. E. Sharpe.

———. 1993. *China's Environmental Crisis: An Inquiry into the Limits of National Development*. Armonk, N.Y.: M. E. Sharpe.

Van Slyke, L. P. 1988. *Yangtze: Nature, History, and the River*. New York: Addison-Wesley Publishing Co., Inc.

People have an insatiable appetite for pandas. But the cult of the giant panda can either help or hurt the conservation of the giant panda. It's up to us to decide.

Wang, D., and S-J. Shen. 1987. *Bamboos of China*. Portland, Ore.: Timber Press.

Wen, H., and Y. He. 1995. The giant panda in Henan, Hubei, Hunan, and Sichuan during the last five thousand years. In *Zhongguo lishi shiqi zhiwu yu dongwu binqian yanjiu*, Wen Huanren et al., eds., 232–239. Chongqing: Chongqing chubanshe. (Pages translated by Mark Elvin.)

CHAPTER 7
THE CONSERVATION CHALLENGE

Dunham, W. 2001. Human activity harms giant panda reserve in China. Reuters News Service, April 6, 2001, at http://www.planetark.org.

Ferraro, P. J. 2001. Global habitat protection: Limitations of development interventions and a role for conservation

Our grandchildren will judge our commitment to saving the Earth.

performance payments. *Conservation Biology* 15:990–1000.

Ferraro, P. J., and R. D. Simpson. 2001. Cost-effective conservation: A review of what works to preserve biodiversity. *Resources* issue 143:17–20.

Janzen, D. 1999. Gardenification of tropical conserved wildlands: Multitasking, multicropping, and multiusers. *Proceedings of the National Academy of Science, USA* 96:5987–5994.

Johnson, K. G., Y. Yao, C. You, S. Yang, and Z. Shen. 1996. Human/carnivore interactions: Conservation and management implications from China. In *Carnivore Behavior, Ecology, and Evolution,* vol. 2, J. L. Gittleman, ed., 337–370. Ithaca, N.Y.: Cornell University Press.

Kaufman, H. 1960. *The Forest Ranger: A Study in Administrative Behavior.* Washington, D.C.: Resources for the Future Press.

Kleiman, D., and M. Roberts, eds. 1991. *Giant Panda and Red Panda Conservation Workshop Working Group Reports.* Washington, D.C.: Smithsonian National Zoological Park.

Li, Z., and Z. Zhou. 1998. Update on activities implementing the National Conservation project for the giant panda and its habitat and current status of giant panda conservation. In *International Workshop on Feasibility of Giant Panda Reintroduction,* S. Mainka and Z. Lu, eds., 209–213. Beijing: China Forestry Press.

Liu, J. 2001. Integrating ecology with human demography, behavior, and socioeconomics: Needs and approaches. *Ecological Modeling* 140:1–8.

Liu, J., Z. Ouyang, W. W. Taylor, R. Groop, Y. Tan, and H. Zhang. 1999. A framework for evaluating the effects of human factors on wildlife habitat: The case of giant pandas. *Conservation Biology* 13:1360–1370.

Liu, J., Z. Ouyang, Y. Tan, J. Yang, and H. Zhang. 1999. Change in human population structure: Implications for biodiversity conservation. *Population and Environment* 21:45–58.

Liu, J., M. Linderman, Z. Ouyang, L. An, J. Yang, and H. Zhang. 2001. Ecological degradation in protected areas: The case of Wolong Nature Reserve for giant pandas. *Science* 292:98–101.

Louchs, C. J., Z. Lu, E. Dinerstein, D. Wang, D. Fu, and H. Wang. 2001. *Conserving Landscapes for Endangered Species: Conservation of the Giant Panda and Its Habitat in the Qinling Mountains.* Washington, D.C.: World Wildlife Fund.

Lu, Z., W. Pan, X. Zhu, D. Wang, and H. Wang. 2000. What has the panda taught us? In *Priorities for the Conservation of Mammalian Diversity: Has the Panda Had Its Day?* A. Entwistle and N. Dunstone, eds., 325–334. New York: Cambridge University Press.

Lu, Z., and E. Kemf. 2001. *Wanted Alive! Giant Pandas in the Wild.* A WWF Species Status Report. Gland, Switzerland: WWF International.

Maple, T. L. 2001. *Saving the Giant Panda.* Atlanta: Longstreet Press.

MacKinnon, J., and R. De Wulf.1994. Designing protected areas for giant panda conservation. In *Mapping the Diversity of Nature,* R. I. Miller, ed., 127–142. London: Chapman and Hill.

MacKinnon, J., F. Bi, M. Qui, C. Fan, H. Wang, S. Yuan, A. Tian, and J. Li. 1989. *National Conservation Management Plan for the Giant Panda and Its Habitat.* Beijing: Ministry of Forestry, and Gland, Switzerland: World Wide Fund for Nature.

Sage, D. 1935. In quest of the giant panda. *Natural History* 35:309–320.

Schaller, G. B. 1998. Giant panda biology and its relevance to reintroduction efforts. In *International Workshop on Feasibility of Giant Panda Reintroduction,* S. Mainka and Z. Lu, eds., 182–184. Beijing: China Forestry Press.

———. 2001. Foreword. In Z. Lu and E. Kemf. 2001. *Wanted Alive! Giant Pandas in the Wild,* 1–3. A WWF Species Status Report. Gland, Switzerland: WWF International.

Shen, M. 2001. How the logging ban affects community forestry management: Aba Prefecture, North Sichuan, China. Document at http://www2.eastwestcenter.org/environment/CBFM/Mao%20Ying.pdf.

Studley, J. 1999. Environmental degradation in southwest China. *China Review* issue 12:1–8.

Zelin, M. 2001. Themes in Chinese history. Document at www.stanford.edu/class/history92a/readings/Zelin.html.

PHOTO CREDITS

Heather Angel: i, 20, 35, 37, 39, 40, 53, 56–57, 65, 70–71, 112, 115, 116, 120–121, 122–123, 132–133, 134, 135, 138, 140, 142–143, 145, 149, 152, 157, 173, 188–189, 194

Erwin and Peggy Bauer: 8–9, 55

Chart on p. 43 adapted from Bininda-Emonds, Gittleman, and Purvis, 1999, Building large trees by combining phylogenetic information: a complete phylogeny of the extant Carnivora (Mammalia).

Jessie Cohen: ii, iv–v, vi–vii, viii–ix, x, xii, xiv–xv, 2–3, 4–5, 6, 10–11, 12, 13, 14–15, 17, 18, 19, 21, 22, 23, 24, 26, 27, 28–29, 30, 32–33, 36, 37, 48–49, 50, 52, 58, 59, 60, 60–61, 62, 63, 66–67, 68, 69, 74–75, 77, 78, 80–81, 83, 90–91, 97, 99, 100–101, 102, 103, 104–105, 106, 107, 108–109, 110, 113, 117, 118, 129, 130–131, 147, 164–165, 166–167, 170–171, 172, 176, 179, 180, 182, 184, 185, 186–187, 195, 196, 197, 198

Jim Messina: 34

John Seidensticker: 72, 73, 92–93, 125, 126, 127, 128, 139, 146, 148, 160, 162, 163, 169, 174, 176, 178

Keren Su: 7, 41, 42, 44–45, 84, 85, 86, 88, 94–95, 96, 124, 136–137, 150, 154–155, 158, 161, 183, 192–193

Art Wolfe: 199

INDEX

Page numbers in **boldface** indicate illustrations